AF251383

Springer Series in
MATERIALS SCIENCE 52

Springer
Berlin
Heidelberg
New York
Hong Kong
London
Milan
Paris
Tokyo

Physics and Astronomy ONLINE LIBRARY

http://www.springer.de/phys/

Springer Series in
MATERIALS SCIENCE

Editors: R. Hull   R. M. Osgood, Jr.   J. Parisi

The Springer Series in Materials Science covers the complete spectrum of materials physics, including fundamental principles, physical properties, materials theory and design. Recognizing the increasing importance of materials science in future device technologies, the book titles in this series reflect the state-of-the-art in understanding and controlling the structure and properties of all important classes of materials.

Series homepage – http://www.springer.de/phys/books/ssms/

Andreas Heilmann

Polymer Films with Embedded Metal Nanoparticles

With 106 Figures

 Springer

Dr. Andreas Heilmann
Fraunhofer-Institut
für Werkstoffmechanik
06120 Halle
Germany

Series Editors:

Professor Robert Hull
University of Virginia, Dept. of Materials Science and Engineering, Thornton Hall
Charlottesville, VA 22903-2442, USA

Professor R. M. Osgood, Jr.
Microelectronics Science Laboratory, Department of Electrical Engineering
Columbia University, Seeley W. Mudd Building, New York, NY 10027, USA

Professor Jürgen Parisi
Universität Oldenburg, Fachbereich Physik, Abt. Energie- und Halbleiterforschung
Carl-von-Ossietzky-Strasse 9-11, 26129 Oldenburg, Germany

ISSN 0933-033x

ISBN 3-540-43151-9 Springer-Verlag Berlin Heidelberg New York

Library of Congress Cataloging-in-Publication Data: Heilmann, Andreas, 1960– Polymer films with embedded metal nanoparticles/ Andreas Heilmann. p. cm.– (Springer series in materials science ; 52) Includes bibliographical references and index. ISBN 3540431519 (alk. paper) 1. Polymers. 2. Thin films. 3. Nanoparticles. 4. Metal Clusters. I. Title. II. Series. TA455.P58 H45 2002 621.3815'2–dc21 2002075896

Springer-Verlag Berlin Heidelberg New York
a member of BertelsmannSpringer Science+Business Media GmbH

http://www.springer.de

Typesetting: Frank Herweg, Leutershausen
Cover concept: eStudio Calamar Steinen
Cover production: *design & production* GmbH, Heidelberg

Printed on acid-free paper   57/3111/ba   5 4 3 2 1

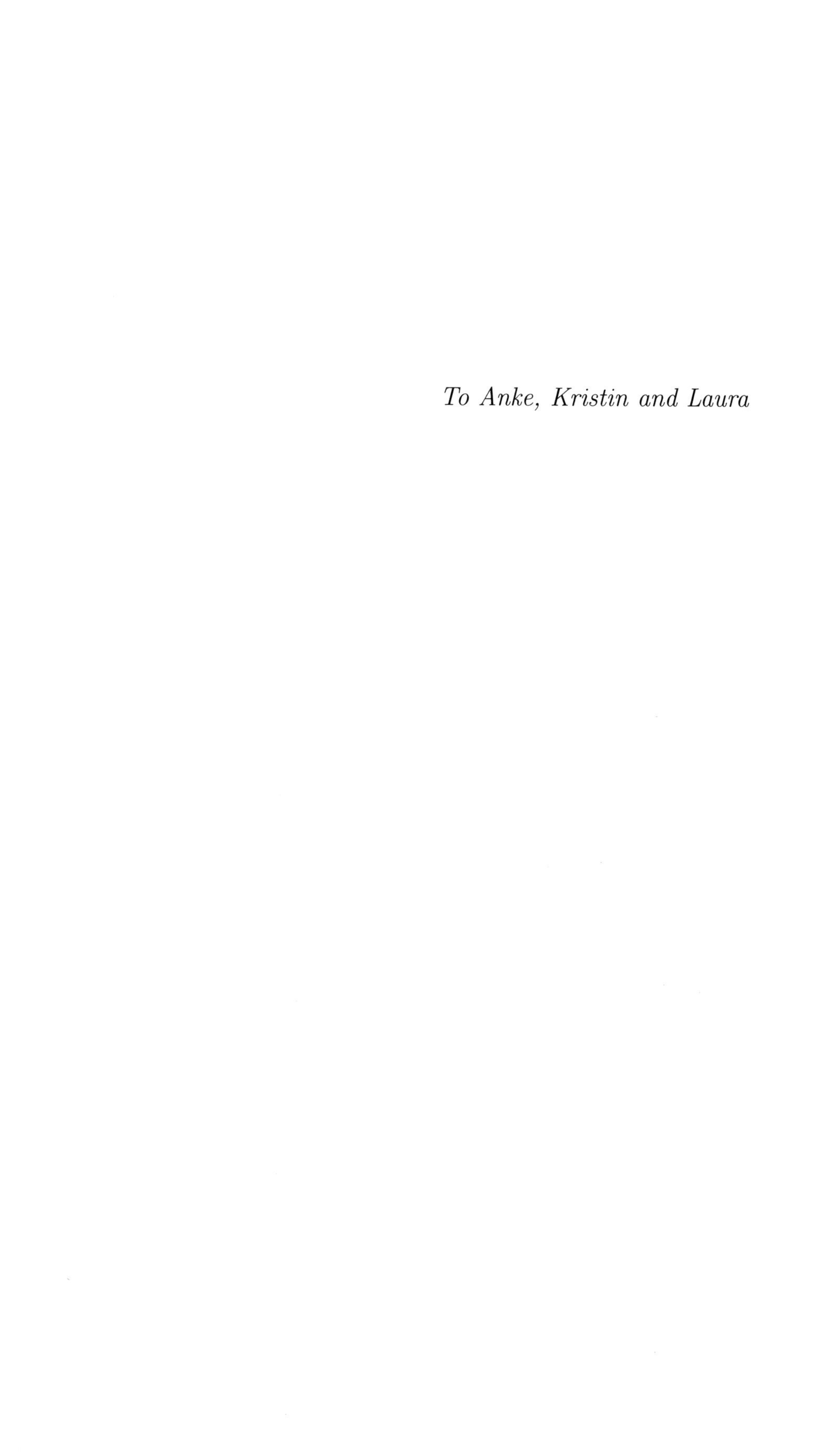

To Anke, Kristin and Laura

Preface

The German physicist and poet Lichtenberg knew nothing about polymer films or nanoparticles. But it is quite likely that he had seen colored glasses with embedded metal nanoparticles made by unidentified Roman glassmakers or so-called ruby glasses created by Johann Kunckel (ca. 1630–1703), a German alchemist. Colored glasses were one of the first nanoparticle-containing materials designed by man. Kunckel did not have the equipment to investigate the structure of the material he had created and could not know that the deep purple color was a result of embedded gold nanoparticles.

These days, however, we have excellent tools such as electron microscopes for the determination of small structures with dimensions in the nanometer range. It is now well-established that such small particles have physical and chemical properties that are very different to those of the bulk material. Indeed an exact knowledge of the nanostructure and of its relation to the physical properties turns out to be essential for the design of nanostructured granular materials and for their industrial application.

It is this requirement that provided the motivation for this book. The aim is to present a detailed study of the correlation between nanostructure and physical (optical, electrical) properties. Due to the great variety of nanoparticle-containing materials, this study can only be performed by focusing on a selected class of materials. These materials are polymer thin films with embedded nanoparticles deposited by vacuum processes. This choice was made for several reasons. Firstly, films prepared in this way display a great variety of very different nanostructures and physical properties. Secondly, the nanostructure and the physical properties can be investigated without extensive sample preparation. Furthermore, these films can be used to investigate the processes taking place during thermal treatment, laser or electron irradiation. All these treatments can result in changes of the nanostructure. A final reason is the enormous application potential of such films in thin film technology. Hence, the aim of this monograph is to describe the detailed determination of the nanostructure of thin polymer films with embedded nanoparticles and its correlation to the optical and electronic properties. The vacuum deposition of polymer films was carried out by a low-temperature chemical vapor deposition process, referred to as plasma polymerization. Incidentally, Lichtenberg was also one of the first scientists to explore plasma processes.

The book was conceived with the intention of making experiments and theoretical considerations easily comprehensible and of allowing the methodological pathways to be easily transferred to other materials with similar nanostructures.

Many individuals have contributed ideas, suggestions, and experimental work which appear in this book. I express my gratitude to Eric Kay and James E. Morris who encouraged me to write this monograph. I wish to acknowledge Claus Hamann, Volkmar Hopfe, Dieter Gerlich, Uwe Kreibig and Dieter Katzer for their encouragement during recent years. I am also grateful to Andreas Kiesow for his experimental contribution and for a critical reading of the manuscript. Further experimental contributions were made by, among others, Jens Werner, Do Ngoc Uan, Wolfgang Grünewald, Michael Quinten, Anne Müller, Falk Müller, Frank Homilius, Mike Gruner, Dirk Schwarzenberg, and Nico Teuscher. Sincere thanks go to them all. Last but not least I thank the Springer editors Claus Ascheron and Angela Lahee for their patience during the preparation of the manuscript.

Augustusburg,   *Andreas Heilmann*
June 2002

Contents

1. Introduction

Composite materials, that is, materials that develop by mixing two or more basic constituents, are a topic of materials science with dramatically increasing interest [1–4]. Applications of fiber composite materials in particular have increased enormously as a result of their mechanical properties. Apart from these materials, applications of nanostructured composite materials, in which one or both of the basic components are structured in the nanometer region, are still in their early stages.

Amongst these nanostructured materials, insulating materials with embedded metal nanoparticles are under focus because of their special structural properties and the extraordinary optical and electrical properties that these confer upon them [5–7]. The nanostructure of such composite materials is determined by the spatial distribution, size and shape of the dispersed metal particles and also by the material properties of the basic constituents themselves. Knowledge of the nanostructure is the key to understanding their macroscopic optical, electrical and mechanical properties.

Beside the relationships between nanostructure and optical and electronic properties, very small metal particles dispersed in an insulating matrix exhibit exceptionally interesting physical properties due to quantum size effects [8–12]. Further, effects are possible for metal particles such as a decrease in melting point [13] or superconductivity although the *bulk* metal does not convert into the superconductive state as at Bi particles [14].

Amongst the wide range of insulating materials with embedded metal nanoparticles, thick or thin insulating layers with embedded nanoparticles have raised special interest. Examples are cermet films (ceramic–metal composite films) or thin polymer films with embedded metal nanoparticles. Polymer thin films are especially suitable as host materials for nanoparticles, whilst their chemical structure and physical properties can be very different. Further, extensive management of the nanostructure is possible during the deposition process of polymer films with embedded nanoparticles. This can be done, for example, by self-organization of colloidal metal particles in polymer solutions or by combination of various vacuum deposition processes.

Up to now, most nanostructural investigations of polymer films with embedded nanoparticles have considered first and foremost only the content of metal particles in the film as a whole. This filling factor quantifies the metal content in the whole composite film and can be specified both as a mass filling factor f_M and a volume filling factor f_V. Since the densities of the metal and the insulator often differ greatly, the volume filling factor of the metal f_V (hereafter denoted f) will be used in most cases. A determination of the filling factor can be made from the simple equation

$$f = \frac{m/V - \rho_{ins}}{\rho_{me} - \rho_{ins}} \, ,$$

where ρ_{me} is the density of the metal and ρ_{ins} is the density of the insulator, for which the volume V of the film (sample area multiplied by film thickness) and the densities of the basic materials must be known. For example, the mass m of the film can be determined by weighing the substrate before and after film deposition.

Depending on the filling factor f, three structure ranges can be distinguished in insulator films with embedded nanoparticles:

- $f \leq f_c$ – insulating structure range,
- $f \approx f_c$ – percolation range,
- $f \geq f_c$ – metallic structure range.

If the filling factor is below the percolation threshold $f \leq f_c$, the metal particles exist separated from each other. The metal-containing film has electrically insulating properties, but the conductivity is significantly higher with embedded metal particles than the conductivity of the pure insulating material. The percolation threshold f_c is defined as the filling factor where the film switches from an insulating to a metallic conductive film. Changes in the d.c. electrical conductivity in the percolation area may amount to eight or more orders of magnitude. In the metallic structure range, particles are no longer completely separated by the insulator matrix, and so the composite film has a higher electrical conductivity, although far below bulk conductivity in a solid state metal.

So far, a uniform distribution of the metal particles in the insulator matrix has been assumed. This becomes less probable as layers become thinner. During film deposition, a laterally and vertically inhomogeneous nanoparticle distribution often develops. It can be observed that to differing degrees the particle distribution does not form in the same way on the interfaces with the substrate and within the film. The actual nanostructure of a real metal-particle-containing insulating film is very complicated and varied, depending on the fabrication parameters and the different possibilities for particle generation.

In the following, this will be discussed by means of examples for the embedding of metal particles into an insulating matrix (Fig. 1.1). A distinction must first be made between lateral and vertical particle distribution. As the film thickness decreases, the ideal case of a laterally and vertically homogeneous nanostructure becomes less probable.

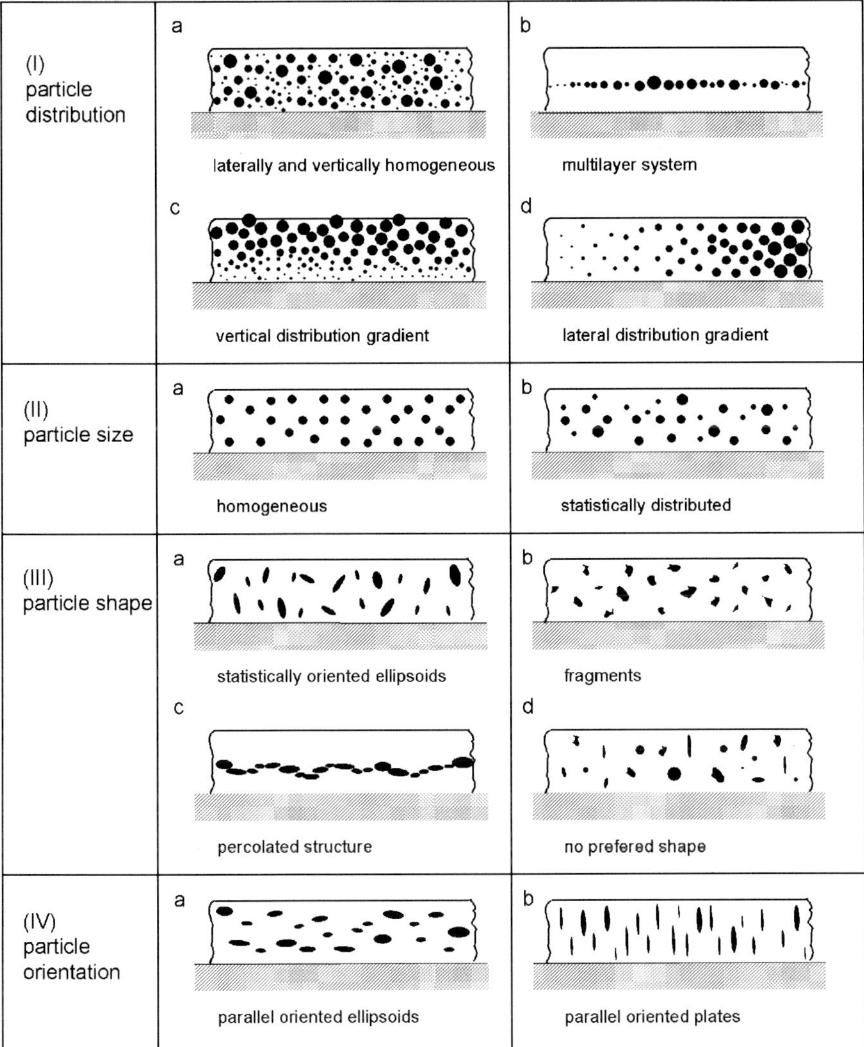

Fig. 1.1. Distributions of particles embedded in another medium

Laterally homogeneous films are often found in which metal particles are practically embedded in one plane, and not on top of each other. In this case it makes sense to call it a multilayer system consisting of a first polymer layer, a polymer layer with embedded metal nanoparticles and a second polymer layer. The filling factor should be used only for the metal-containing layer, considering that the optical properties are determined only by the size and number of the embedded particles and not by the thickness of the embedding, a mostly transparent polymer matrix.

A vertical distribution gradient of the metal particles can be generated by an appropriate choice of fabrication conditions. This film nanostructure is used, for example, for the metalization of polymeric materials in order to resolve the large macroscopic interface between metal and polymer into many small microscopic interfaces, and thereby increase the adhesive strength of the subsequently deposited metal film. Here, a determination of the filling factor is only possible for the whole film and is not actually meaningful.

Metal-containing polymer films with a lateral particle distribution gradient have a continuously changing metal content over a larger substrate area. Therefore, the size and shape of the embedded metal particles change as the metal content increases. Experimentally determined filling factors can only give an average value for a defined area of the film. Metal-containing polymer films with a continuously changing metal content represent an ideal subject of investigation when determining the influence of different particle sizes and shapes on the optical and electrical properties of composite films.

Apart from the spatial distribution of the metal particles in the polymer film, particle size is also an important parameter. Particles can possess either an almost uniform size or else be subjected to a statistical size distribution. For the investigation in particular of the collective optical properties of the metal particles, a uniform particle size is often demanded, but is seldom realized experimentally.

Furthermore, the shape of the particles greatly influences physical properties. So far, only spherical particles have been studied. However, particles must also be considered that do not bear the slightest resemblance to a sphere. Particles with the shape of rotationally symmetric ellipsoids can be embedded, as can non-rotationally symmetric, fragment-like particles. For ellipsoidal particles, the ratio between major and minor axes can be constant or subject to statistical distribution. In addition, the different orientations of the ellipsoids with respect to the substrate must be considered. A preferred direction may or may not exist. This is very important for the physical properties of films with very long elongated spheroidal particles.

The complexity involved in describing nanostructures is demonstrated by comparing various percolation structures (Fig. 1.1 IIIc) with those formed by parallel oriented ellipsoids (Fig. 1.1 IVa). Both films have the same filling factor in relation to the total film thickness, but a different vertical particle distribution. This leads to few differences in the optical properties, but very different electrical transport properties. There are only small spacings between the particles and a percolation structure with the formation of conducting paths can be observed in Fig. 1.1 IIIc. The electrical d.c. conductivity is much higher for a film with this percolation structure than for a film without it.

Depending on the various types of metal embedding that have been discussed, it can be seen that the filling factor f is only meaningfully applicable to describe the nanostructure if polymer films with embedded metal nanoparticles have a homogeneous lateral and vertical particle distribution. For all other nanostructures, the filling factor seems to be rather irrelevant, and a simple determination of the filling factor with the weighing method described previously does not yield any information suitable for describing relationships between nanostructure and physical properties. Extensive nanostructural analysis is needed in order to determine the lateral and vertical particle size and shape distribution.

A large number of modern analytical techniques are available for determining the nanostructure of polymer thin films with embedded nanoparticles. Transmission electron microscopy (TEM) is the most appropriate and powerful amongst these. However, it is not sufficient to carry out simple lateral investigations of the nanostructure. It is also necessary to investigate the vertical nanostructure with the help of a cross-sectional sample preparation. The differences in the nanostructure between the film with a percolation structure and that with embedded parallel oriented ellipsoids in Figs. 1.1 IIIc and IVa are only visible in the vertical view, whereas the lateral electron microscope images look almost identical. It is also useful to support TEM nanostructure investigations with at least one additional analytic method such as photoelectron spectroscopy (XPS) or Auger electron spectroscopy (AES).

Up to now, a time constant and fixed nanostructure has been assumed for the polymer film with embedded nanoparticles after it is deposited. However, possible modifications in the nanostructure must be considered, for example, due to thermal treatment. It is obvious that small changes in the particle shape and size distribution close to the percolation threshold can generate significant changes in electrical d.c. conductivity.

In contrast to ceramic or oxide thin film matrices like Al_2O_3, MgO or SiO_2, polymer matrices possess weaker thermal stability. However, modifications in the nanostructure due to laser irradiation or electron beam irradiation can be much more easily realized and offer a great variety of possibilities for position-resolved nanostructural modifications and artificial nanostructures.

When changes occur in the nanostructure, it is necessary to study the particles themselves, the interface between matrix and particle, and the atomic diffusion of the metal through the matrix. Nanostructural changes can result in shape modifications due to recrystallization or in the unification of two neighboring particles (coalescence). Single particle migration or the formation of chemically modified shells around the metal particle are also observed.

For these investigations of nanostructural changes, very thin polymer films are particularly suitable due to their transparency in TEM. For fabrication of the thin polymer matrix, a vacuum deposition technology called plasma polymerization was therefore selected.

The use of plasma polymerization for the fabrication of thin polymer films is an established, low-pollutant method which is compatible with other thin film technologies, especially in microelectronics. Consequently, this deposition method has an increasing number of applications. Metal particles are embedded during deposition of the plasma polymer film, mainly by simultaneous metal evaporation or metal sputtering. Low-thickness plasma polymer films (< 200 nm) are mainly optically transparent in the visible spectral region, and they are also transparent in TEM. Hence, films with embedded metal particles can be observed directly in TEM without further thinning. Another advantage of plasma polymer films is that their chemical structure and hence also their electrical and optical properties can be varied by changing deposition conditions.

Through combination of two vacuum deposition technologies, polymer films with embedded nanoparticles can be fabricated with a very large variety of nanostructures. This begins with laterally and vertically homogeneous metal-containing films and continues with multilayer systems and metal-containing films with lateral and vertical particle distribution gradients (see Fig. 1.1). In particular, films with a lateral particle distribution gradient enable a comprehensive correlation of nanostructure with electrical and optical properties for the different particle size and shape distributions and uniform matrix properties.

To study the relationships between nanostructure and nanostructural changes on the one hand and optical and electrical properties on the other, this book focuses on plasma polymer thin films which are an ideal matrix for embedded nanoparticles. Investigations concentrate on metal nanoparticles of silver, gold and indium. The optical properties of silver and gold particles are especially interesting because the plasma resonance absorption in the visible region depends strongly on particle size and shape distribution. Indium particles have been selected because they present a large number of possibilities for generating nanostructural changes.

Because of the very complex interface properties of metal nanoparticles in self-organized nanoparticle assemblies and also in order to focus discussions of the nanostructure–property relationship, self-organized nanoparticles embedded in polymer matrices will be considered for comparison. Self-organized polymer films containing metal nanoparticles, as well as the chemical properties of metal clusters are reviewed in [15–18], for example.

The book is organized as follows. Film fabrication is described in Chap. 2, where emphasis is placed on characterising the plasma polymer matrix material. The fabrication of plasma polymer thin films with embedded metal nanoparticles by metal sputtering or by metal evaporation during plasma polymerization is explained in detail.

Chapter 3 describes characterization of the nanostructure of plasma polymer films with embedded metal nanoparticles using transmission electron microscopy and photoelectron spectroscopy. Here the main point is to determine particle size and shape distributions in the lateral and vertical directions.

According to the description of possible physical processes that lead to changes in the nanostructure, the results of nanostructural changes in plasma polymer multilayers with embedded nanoparticles caused by thermal treatment, laser irradiation and electron irradiation are given in Chap. 4. Nanostructural modifications which are spatially resolved to the order of magnitude of the embedded metal particles are discussed, especially those studied through in situ observations.

In Chap. 5, the electrical d.c. conductivity of plasma polymer thin films with embedded nanoparticles is discussed. Chapter 6 gives a comprehensive description of experimental optical properties in the UV–visible–NIR spectral region for films before and after nanostructural changes. Optical properties calculated using optical scattering theories and various effective medium theories are presented.

2. Film Deposition

2.1 Plasma Polymer Thin Films

2.1.1 Organic Thin Films

The large number of applications of thin films based on polymerized hydrocarbons has led to many methods for producing this kind of film. These deposition technologies can be classified in different ways:

- By using the physical state of the starting material:
  - solid state → gaseous state → film,
  - solution → film,
  - gaseous state → film.
- By characterizing the environment during deposition:
  - ultra high vacuum,
  - pre-vacuum,
  - atmospheric pressure.
- By using the movement of the substrate during deposition:
  - dip coating,
  - spin coating,
  - roll deposition,
  - stationary substrate.
- By the film forming process which is the most common method for deposition processes starting from the gaseous state:
  - Chemical vapor deposition (CVD):
    - plasma assisted CVD (PACVD),
    - laser assisted CVD (LACVD),
    - photochemical vapor deposition.
  - Physical vapor deposition(PVD):
    - evaporation,
    - sputtering,
    - cluster beam deposition.

Further classifications are possible, for example, using the polymerization temperature, the polymerization method, or the polymerization product. These different conditions of fabrication yield thin polymer films with widely ranging properties such as:

adhesive strength: perfect adhesion properties $\leftrightarrow$ negligible adhesion,

surface tension: hydrophobic $\leftrightarrow$ hydrophilic,

electrical conductivity: highly insulating $\leftrightarrow$ semiconducting (conducting),

optical behavior: transparent $\leftrightarrow$ almost completely absorbing.

Plasma polymerization is a method for plasma assisted chemical vapor deposition of a gaseous basic material (monomer) which is polymerized during deposition. The thin polymer film grows with the assistance of an electrical gas discharge. The layer-forming process usually takes place in vacuum at a total pressure in the range 10^{-1}–10^3 Pa and at temperatures between 290 K and 370 K. Apart from the need to clean a stationary or moving substrate, special preparations are not required.

In principle, so-called amorphous hydrogenated carbon (a-C:H) films or amorphous silicon hydrocarbon (a-C,Si:H) films can be deposited in the same way, but the energy impact for the deposition process is much higher. The transition between plasma polymer films and a-C:H or a-C,Si:H films is gradual [19], but the deposition of a-C:H films usually aims to give films with high hardness (hard coatings) or wear resistance. In the following, the films described are designed as plasma polymers but some of their properties, like adhesion, are similar to those of a-C:H films. The deposition of a-C:H and also a-C,Si:H films is described in e.g. [20–22].

2.1.2 Plasma Polymerization

The aim in the following brief commentary on plasma polymerization will be to understand the basic physical properties of films deposited in this way. Comprehensive reviews of plasma polymerisation, plasma surface treatment and plasma deposition are given in the books by Yasuda [23], d'Agostino [24], Inagaki [25] and review articles [26, 27]. The concepts underlying the gaseous state processes are extensively described in [28–30] among others.

As a method for fabricating thin polymer coatings, plasma polymerization can generally be considered as polymerization of an ionized monomer in a gaseous state by collisions with accelerated electrons in the plasma caused by an electrical discharge. The fundamental components of a reactor for plasma polymerization are a vacuum system, a plasma generator, a controllable gas inlet, and a system for in situ analysis, e.g., thickness monitoring or optical emission spectroscopy.

Because the electrical discharge usually takes place in vacuum at pressures of 1–100 Pa, the reactor (glass, quartz, high-grade steel) is evacuated by mechanical vacuum pumps, but sometimes also by additional turbo molecular pumps. In principle, plasma polymerization is also possible under normal atmospheric pressure conditions (see [31]).

Inside or outside the reactor, the gaseous or liquid monomer is often mixed with a predetermined amount of an inert gas such as argon. In a closed system, the supply of new monomer is stopped before or during the plasma polymerization. Only gas remaining in the reactor can take part in polymerization. In an open system, the most common case, a constant supply of monomer gas is provided. Furthermore, an additional reactive gas such as oxygen can be introduced.

Two electrodes are usually located inside the reactor. The glow discharge takes place between these electrodes. If reactors with metallic reactor walls are used, one single electrode is sufficient because the walls act as the second electrode. The glow discharge can be produced by a high voltage transformer (a.c. 50 Hz), by a radio frequency generator (r.f. 13.56 MHz) or by a microwave generator (2.45 GHz). Since at 50 Hz, the ionized monomer fragments are able to follow the periodic field changes, the 50 Hz alternating current (a.c.) discharge acts as a pulsed direct current (d.c.) discharge [32].

According to Kay [29], the electron density n_e and energy distribution $f(E)$ which describe an atomic gas plasma can be used to describe the polymerization plasma. Both of these quantities are determined by the kind of monomer, the frequency and electrical power of the plasma generator, the monomer flow rate, the pumping behavior of the vacuum pump, and the reactor dimensions. Increasing the electrical power raises the electron density n_e and the production rate of active chemical compounds increases. Values of the electron density n_e and the energy distribution $f(E)$ are difficult to determine, but are very useful for comparing different plasma discharges or for technical upscaling [29]. Poor knowledge of real polymerization conditions is often described as a so-called reactor dependence of film deposition. In fact, details of plasma polymerization with a given monomer are not immediately transferable to plasma polymerization of the same monomer in a reactor of different dimensions. One possible way to avoid this reactor dependence could be to use an idealized normal reactor, but this would only be meaningful for research into the basics of film deposition processes.

As suggested in [33], reactive and non-reactive reaction products in the plasma polymer originate from the monomer. The reactive products also participate in forming the plasma polymer, where they produce further non-reactive products. The plasma etching process must also be considered, because this process also generates non-reactive products. These individual reaction channels are modelled by a set of differential equations [30, 33].

According to Yasuda [23], two different plasma polymerization conditions can be distinguished. With sufficient monomer supply, the electrical energy introduced is not high enough to ionize the monomer gas completely. The deposition rate grows linearly with the quotient P/FM of the electrical power P and the mass flow rate FM, where F is the flow rate and M the molecular weight of the monomer. This plasma polymerization regime is said to be low-energy. Using higher discharge power the energy introduced is enough to ionize the monomer gas completely and the deposition rate no longer increases with the electrical power. This plasma polymerization is monomerless. Since most bonds of the monomer are broken, the original structure of the monomer is completely destroyed. A certain value of P/FM can be given at which the transition between these two polymerization conditions takes place [23].

The layer forming process during plasma polymerization takes place mainly on the electrodes, but also on the reactor walls and on all parts inside the reactor. The formation of powders and oils is also possible, but is inhibited by an appropriate choice of fabrication parameters, especially the monomer flow rate. Deposition rates depend on the monomer supplied and the polymerization parameters and can be adjusted over a wide range of values, viz.,

$$\frac{\delta d}{\delta t_{\mathrm{pp}}} = 1\text{--}100 \ \mathrm{nm\,min}^{-1} .$$

The deposition rate is influenced by the nature of the substrate (electrically conductive or insulating) and also by the position of the substrate in the reactor (on the electrodes or on the reactor walls) and decreases with increasing substrate temperature [23, 34, 35]. Since a plasma etching process occurs at higher plasma powers, each plasma polymerization reactor has its own maximum deposition rate [37] for a certain constant monomer flow rate.

The electrostatic charge during the plasma discharge (biasing) of the electrodes influences the plasma polymerization process. By applying an additional external bias voltage, the energy of positive ions in the vicinity of the electrodes can be varied and so therefore can the deposition rate [36].

Early investigations mainly concerned the plasma polymerization of fluorine hydrocarbons because of the extraordinary properties of fluorine hydrocarbon polymers [29]. The ecological need to substitute for the fluorine- and chlorine-containing polymers means that silicon organic monomers and hydrogenated hydrocarbons (alkanes, alkenes, aromatic compounds) are now being used. The most frequently used monomers are hexamethyldisilazane (HMDSN) and hexamethyldisiloxane (HMDSO). Furthermore, monomers with special functional groups or conductive organic compounds such as thiophene [38–40] and metal organic compounds [41] can also be plasma polymerized.

The chemical structures of plasma polymer layers are characterized by chaotic, branched and crosslinked polymer chains of inhomogeneous lengths, with open bonds and a certain amount of free radicals [32]. If at higher power densities the plasma polymerization becomes monomerless, the chemical structure of the monomer is less and less relevant for the later structure of the plasma polymer. For plasma polymer films made from simple monomers with only one monomer group (hydrogenated hydrocarbons, fluorinated hydrocarbons or silicon organic compounds), the chemical structure can seldom be described using typical quantities like chain length or degree of polymerization. For plasma polymerization of a monomer with a special functional group, polymerization conditions can be chosen in such a way as to maintain these functional groups. It is thus possible, for instance, to produce photochromic layers with similar reversible photo-isomerization to that of the monomer [42].

The thermal stability of plasma polymer films is much higher than for polymer materials made by conventional polymerization using the same monomer. Thermo-gravimetric measurements or differential calorimetric measurements show thermal stability (mass loss $\leq 5\%$) of plasma polymer layers made from perfluorine benzene up to 626 K [43] or from HMDSN up to 570 K [44], and even up to 700 K [45].

From the applications standpoint, plasma polymer films are very interesting for the functional deposition of polymer materials. Plasma polymer films provide a continuous, pinhole-free layer even on rough surfaces and exhibit very uniform film thickness, properties which cannot easily be achieved using other deposition techniques for polymer thin films.

Plasma polymer thin films are thus used for many applications which can only be described briefly here. Due to their large variety of chemical and physical properties, these films have been proposed for very different applications. To begin with, there are a large number of proposed and realised applications using the corrosion protection or moisture protection properties of plasma polymer films as a barrier or passivation layer [46–60]. Organosilicon monomers like hexamethyldisilazane (HMDSN), hexamethyldisiloxane (HMDSO), vinyltrimethylsilane (VTMS), tetramethylsilane (TMS) or various other silanes, but also ethanes and ethenes as well as fluorocarbons like perfluorobenzene were the main components used. Efforts have also been made to use plasma polymer films for hydrophilisation of cotton fabrics [61]. Otherwise, because of the strong dependence on the deposition properties of films made from these and other monomers, it has been reported that they can also be used as gas separation layers or gas permeation layers [43,62–72]. The surface biocompatibility of some implants can be improved with plasma polymer layers made from ethylene or acrylic acid [73–75]. Biocompatibility of plasma polymer films can also be achieved and the films can be used to prepare substrates for enzyme immobilization, or to cover immobilized enzymes in biosensors [76–78]. The hydrophilic properties of plasma polymer films made from ethylenes or organosilicon monomers have been used for applications in moisture sensors [79–84] and, if special functional groups are placed on the surface, for gas sensors [85–87]. Plasma polymer films can also act as an intermediate layer to improve the adhesion of other layers [88,89].

The plasma polymer deposition process can be very easily introduced as a process step in microelectronics fabrication. Plasma polymer films, mainly deposited from methylmethacrylate, styrene or silicon organic monomers, are proposed as photoresists [90–92], electron beam resists [93–97] or resists for X-ray lithography [98], but also as films for gate insulators [99], intermediate layers [92] or surface passivation layers [100–103]. Because of their weak absorption in the ultraviolet region, very low absorption in the visible spectral region and low light scattering, plasma polymer films, again mainly made from organosilicon monomers, have been proposed and used as UV protection layers [104], antireflection coatings [105], interference layers [106] or in optical wave guides [107, 108].

Many other applications of plasma polymers made from films with various special functional groups have been proposed but they will not discussed in detail here. Every year a large number of publications come out on these topics. The main subject of investigation is the dependence on fabrication conditions (e.g., reactor, monomer) with regard to special applications.

A great many special technical solutions are used for industrial applications. For example, plasma polymer films have been used for surface passivation of metallized car reflectors or lampshades. Due to the simple equipment and intensive optimization of the deposition process which is needed to realize the required functions, deposition process parameters and other cases of special applications are generally strictly confidential.

The variable fabrication conditions, excellent mechanical properties, low optical absorption and low electrical conductivity all contrive to make plasma polymer films an appropriate matrix material for embedding metal nanoparticles.

2.1.3 Fabrication and Properties of Thin Plasma Polymer Films as a Matrix Material for Nanoparticles

Deposition Reactor, Monomer and Substrate. For the investigation presented in the following, plasma polymerization was carried out in high vacuum systems with 30 l glass or metal reactors. The reactors were constructed in such a way that metal evaporation or metal sputtering could occur concurrently with plasma polymerization. Two parallel electrodes with diameters of 160 mm or 150 mm were placed inside the reactors. An electrode spacing of 100 mm was chosen to realize metal evaporation between the electrodes in additional to plasma polymerization (see Fig. 2.10). For simplicity, most of the films reported in the following were deposited by this reactor. In the case of simultaneous metal sputtering, the upper source acts as a second electrode. Before each deposition, the metal plate electrodes are wrapped in aluminum foil to guarantee reproducible discharge conditions.

The 50 Hz a.c. discharges with voltages in the range 500–3400 V were established between the two parallel glow electrodes. The active plasma polymerization power inside the reactor was determined from the primary electrical power supplied, with the help of a calibrating graph. The power consumed by plasma polymerization is 50–90% of the primary power. Measurement of the primary power ensures a sufficient reproducibility of layers, provided that other deposition parameters (monomer flow rate, monomer partial pressure) can be held constant. The primary power P was varied in the various plasma polymerization reactors between $6 \leq P \leq 1000$ W. The power density of plasma polymerization p calculated using the calibrating graph and the area of the electrodes lies in the range $0.03 \leq p \leq 5.00$ W cm^{-2}.

The following monomers were chosen for plasma polymerization:

- benzene C_6H_6,
- styrene $C_6H_5CH–CH_2$,
- hexamethyldisilazane $Si(CH_3)_3–NH–Si(CH_3)_3$ (HMDSN),
- hexamethyldisiloxane $Si(CH_3)_3–O–Si(CH_3)_3$ (HMDSO).

The influence of the specific monomer on the film structure decreases at higher power densities, and the layers made from benzene or styrene and HMDSN or HMDSO become very similar. With further increase in power density, the transition to producing amorphous a:C,H films or a:C,Si,H films is possible, in principle.

Since the simultaneous metal evaporation should be carried out without changing fabrication parameters, a carrier gas was not used and the system was worked with a lower monomer pressure. Under low pressure, the liquid monomer in a glass bulb goes into the gaseous state and flows directly into the reactor. The flow rate was controlled by needle valves or mass flow controllers. It was also obtained from the mass loss of the monomer liquid over very long flow times. After about 500 s, the monomer flow rate was adjusted. The monomer flow rate can be adjusted over 0.01–0.5 $Pa\,l\,s^{-1}$. The absolute pressure before starting plasma polymerization (between 10–100 Pa) is close to the partial pressure of the monomer and depends on the flow rate.

Plasma polymerization conditions for the monomers benzene, styrene, HMDSN and HMDSO are described by the following parameters:

- monomer flow rate:
  - benzene, styrene   0.05 $Pa\,l\,s^{-1}$,
  - HMDSN, HMDSO   0.1 $Pa\,l\,s^{-1}$,
- monomer partial pressure:
  - benzene, styrene   0.08 Pa,
  - HMDSN, HMDSO   0.2 Pa,
- plasma polymerization power density p in $W\,cm^{-2}$,
- plasma polymerization time t_{pp} in s.

Concerning other conditions for further investigation of the films, different substrate materials were used. Glass substrates were used for measuring film thickness after deposition, thermally oxidized (500 nm SiO_2) silicon wafers for measuring adhesion strength and for determining XPS depth profiles, quartz substrates for UV–visible–NIR spectroscopy, KBr crystals for IR spectroscopy and thermally oxidized silicon wafers with metal electrodes for electrical measurements.

Film Thickness. Thickness determination was carried out in situ using a quartz sensor microbalance during deposition or ex situ after film growth. The microbalance measurement is common but has the disadvantage of monitoring the growth of a quartz crystal. The growth behavior at the substrate can be very different from that at the quartz sensor because of shielding due to the quartz oscillator support and different surface conductivities, if aluminum is coated, for example. For this reason, microbalance results have to be confirmed by an additional method for thickness determination.

There are various methods for ex situ determination of thickness after deposition. The use of an interferometric method, an optical distance difference measurement (laser microfocus method) or atomic force microcopy (AFM) requires sharp edges between deposited and non-deposited substrate areas, otherwise the layer must be scratched. For optical methods, a thin reflective metal layer (aluminum) must be deposited. Measurements are also influenced by the surface roughness of the substrate. In the following, film thickness is always rounded to steps of $d \pm 5$ nm.

A very accurate method for determining the thickness of such layers is TEM investigation of cross-sectioned samples (Sect. 3.2.2). Moreover, if the film density is exactly known, the thickness can be determined via a weighing technique. In principle, determinations of film thickness are possible from X-ray photoelectron spectroscopy and infrared (IR) spectroscopy. A further possibility exists using comparisons of optical UV spectra (Sect. 2.1.3).

As described in Sect. 2.1.2, film thickness depends on plasma polymerization power, monomer flow rate, monomer pressure and the kind of substrate, as well as on reactor dimensions.

The left-hand diagram of Fig. 2.1 shows the film thickness d of plasma polymer films made from HMDSN versus the power density p for different plasma polymerization times t_{pp}. It can be observed that the thickness of the layer does not increase after a critical power density is reached at $p \approx 0.3$ W cm^{-2}, and even declines with further increase in power density. This results from the transition from monomer-rich to monomerless polymer-

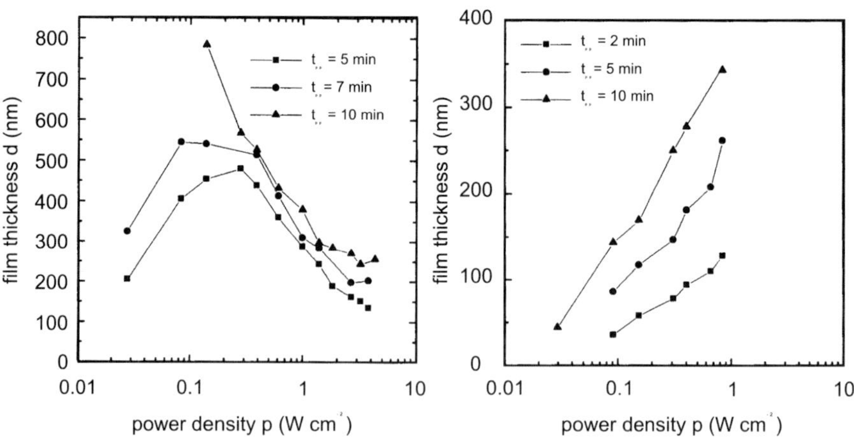

Fig. 2.1. Film thickness d of plasma polymer films as a function of power density p for different plasma polymerization times t_{pp}. *Left*: HMDSN. *Right*: benzene. Substrate: thermally oxidized silicon wafer

ization [23] (see Sect. 2.1.2). In [35], a maximum growth thickness was found at about $p \approx 0.4$ W cm^{-2} in a 13.6 MHz r.f. discharge. This was attributed to film thickness loss due to plasma etching. One further reason for the decrease in deposition rate could be the temperature rise of the electrodes observed for higher powers and longer polymerization times if the electrodes are not cooled. In addition, the chemical structure of the plasma polymer films varies as a function of the power density (see IR investigations in Sect. 2.1.3).

The film thickness of plasma polymer films made from benzene presented in the right-hand picture of Fig. 2.1 increases with the power density p and plasma polymerization time t_{pp}. In contrast to the HMDSN plasma polymerization case, a critical power density was not reached here.

With the quartz sensor microbalance, the thickness growth rate $\Delta d/\Delta t_{pp}$ also changes with deposition time. For instance, the deposition rate of plasma polymers made from benzene at $p = 0.85$ W cm^{-2} falls during the first 60 s of deposition from $\Delta d/\Delta t_{pp} \approx 50$ nm min^{-1} to $\Delta d/\Delta t_{pp} \leq 10$ nm min^{-1} after $t_{pp} = 10$ min. One reason for a decrease in deposition rate with time is that the former conducting polymerization electrodes wrapped in aluminum foil get an insulating layer as a result of the deposition. This in turn modifies discharge conditions inside the reactor. Furthermore, the deposition rate decreases with increasing substrate temperature. For example, it decreases to half the value of plasma polymerization at room temperature for silane plasma polymerization with a substrate temperature of 370 K [34]. High deposition rates from $\Delta d/\Delta t_{pp} \approx 120$–$1200$ nm min^{-1} [51] were realized with HMDSO plasma polymerization.

Mechanical Properties and Density. The adhesive strength of plasma polymer films deposited on silicon wafers has been determined by the very simple scotch tape test. This test offers a scale with five adhesive strength ranges, where $H_s = 0$ means no adhesion and $H_s = 5$ perfect adhesion. Films made from HMDSN and HMDSO with thickness $d \leq 1\,000$ nm show adhesive values of $H_s = 3$–5, whereas films made at higher power densities provided an adhesive value of $H_s = 5$. It is observed that thicker layers show delamination behaviour caused by the inherent tensile stress of the material. This effect is also influenced by moisture. Such patterned delamination structures are also observed on other thin films such as amorphous carbon layers [109]. Plasma polymerized benzene and styrene films exhibit lower adhesive strengths but demonstrate complete adhesion with $H_s = 5$ for layers under $d = 200$ nm. These layers and also those made from HMDSN and HMDSO are not removable by the usual solvents.

Extensive measurements of the adhesion of plasma polymer films (in particular, HMDSN) on PTFE showed adhesive strengths in the range 20–60 kg cm^{-2} [110] or 6–10 kg cm^{-2} for plasma polymerized silane films on titanium substrates. With the nanoindentation method, a hardness of $H_m = 0.3$ GPa and a modulus of elasticity $Y_m = 6$ GPa were determined for plasma polymer films made from C_3F_8 [111]. Plasma polymer films made from HMDSN showed a range of modulus values $Y_m \approx 5\text{–}45$ GPa [112], where these high values are related to a:Si:C:N:H layers.

It is difficult to measure the microhardness of thin polymer films owing to substrate influence. Plasma polymer films were deposited on aluminum bond pads with a natural aluminum oxide layer of 5 nm placed on a silicon wafer. On the basis of load/unload curves measured using registrating hard measurements with an ultra microhardness unit and comparing with results obtained for thin aluminum oxide films on the same substrate, the universal hardness of plasma polymer films deposited from HMDSN was determined as HU $= 1.5\text{–}3.5$ GPa [113]. This is only one order of magnitude lower than for a-C:H thin films.

The density of plasma polymer films made from HMDSN was determined as $\rho \approx 1.3\text{–}1.8\,\mathrm{g\,cm^{-3}}$ [112], where the density increases with increasing power density. Density measurements of plasma polymer films made from HMDSO with similar polymerization power density ranged over $\rho \approx 0.9\text{–}1.6\,\mathrm{g\,cm^{-3}}$ [114]. This is considered to result from the decreasing hydrogen and carbon content in the layers. Other results for the density of plasma polymer films made from HMDSN and HMDSO are given as $\rho \approx 1.2\text{–}1.8\,\mathrm{g\,cm^{-3}}$ or $\rho \approx 1.0\text{–}1.3\,\mathrm{g\,cm^{-3}}$, respectively, in [62]. For amorphous carbon films made from benzene, densities of $\rho \approx 1.5\text{–}1.8\,\mathrm{g\,cm^{-3}}$ have been determined [115].

The density of plasma polymer films can be determined more exactly if thickness measurements made by the quartz oscillation method are combined with another method for thickness determination such as atomic force microscopy (AFM). Comparing the thickness values determined by quartz oscillation and AFM measurements on the quartz oscillator as well as conventional weighing techniques, the density of HMDSN films deposited at $p = 0.4$ W cm^{-2} was determined as $\rho = 1.7 \pm 0.1\,\mathrm{g\,cm^{-3}}$.

Chemical Structure. Infrared (IR) spectroscopy yields little information concerning the chemical structure of plasma polymer films. The very different polymerization grades and varying chain lengths result in IR spectra with very spread out oscillations, and these are rather difficult to interpret.

Figure 2.2 presents the IR transmission spectra of six plasma polymer films, deposited from HMDSO with different power densities, in a wave number range of $1\,300\text{–}600$ cm^{-1}. The measurements were made on films deposited on KBr crystals. Four different peaks can be identified: Si–CH$_3$ deformation oscillation at $1\,260$ cm^{-1}, Si–O–Si stretching oscillation at $1\,040$ cm^{-1} Si–(CH$_3$)$_3$ stretching oscillation at 840 cm^{-1} and Si–(CH$_3$)$_2$ stretching oscillation at 800 cm^{-1}. While clear oscillations are still visible for the smallest power density ($p = 0.14$ W cm^{-2}), it is almost impossible to identify the very broad resonances at higher power densities, or to determine shifts. Since the layers were produced with the same polymerization time ($t_{pp} = 600$ s), the lower thickness of the films deposited with higher power makes the discussion more difficult. But the disappearance of the Si–CH$_3$ deformation oscillation at $1\,260$ cm^{-1} for higher power densities is not caused by decreasing film thickness. Furthermore, the 840 cm^{-1}/800 cm^{-1} double peak of the Si–CH$_3$ stretching oscillation disappears with increasing power density. This decrease in the CH$_3$ proportion shows that the layers which were deposited with increased power density are nearly free of CH groups, and that the chemical structure comes closer to the structure of amorphous a-Si,C:H films. The oxygen proportion represented by the Si–O–Si stretching oscillation at 1 040 cm^{-1} decreases together with the Si–CH$_3$ deformation oscillation. After a thermal treatment of HMDSO films up to 580 K, no spectral shifts were found except for small changes of intensity.

For the main part, IR spectroscopy investigations carried out by other authors [63, 68, 116–122] on film plasma polymerized from HMDSO have produced results similar to Fig. 2.2a. A very similar change of HMDSO plasma polymer films with increasing r.f. power has been observed in [51] and interpreted as a decrease in the organic proportion. The Si–CH$_3$ deformation oscillation at 1 260 cm^{-1} disappears in [51] with an r.f. power

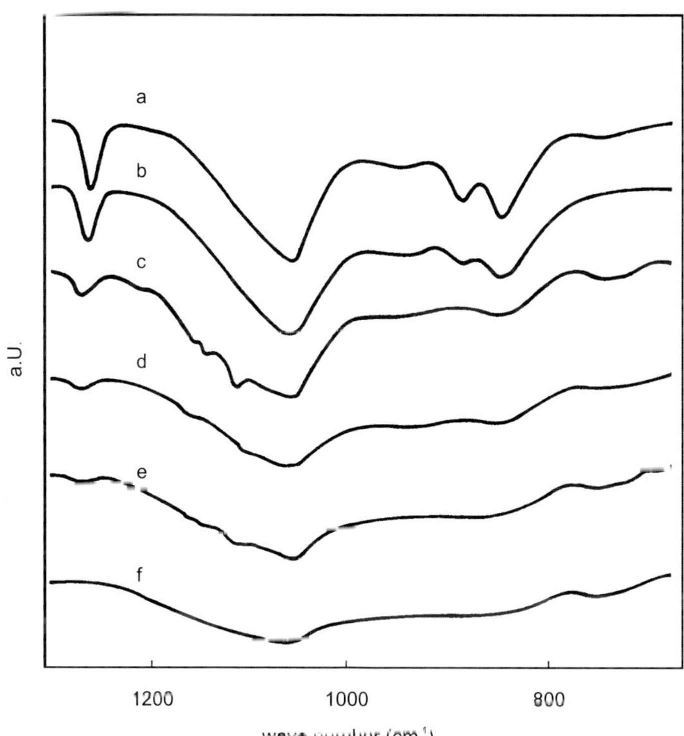

Fig. 2.2. IR transmission spectra of 6 plasma polymer films made from HMDSO at different power densities: (a) $p = 0.14$ W cm^{-2}, $d = 780$ nm, (b) $p = 0.40$ W cm^{-2}, $d = 530$ nm, (c) $p = 1.00$ W cm^{-2}, $d = 380$ nm, (d) $p = 1.88$ W cm^{-2}, $d = 280$ nm, (e) $p = 3.00$ W cm^{-2}, $d = 240$ nm, (f) $p = 5.00$ W cm^{-2}, $d = 250$ nm

of 800 W, which is equivalent to $p = 4.5$ W cm^{-1} within the reactor dimensions described before. A decrease in carbon proportion with increasing power density was also observed in [112]. Longer term storage of the films leads to an indiffusion of oxygen and to higher proportions of OH–, C=O and Si–O–Si groups [123]. With thermal treatment (620 K) of plasma polymer films made from hexamethylcyclotrisilazane, the proportion of hydrogen decreases significantly [124], and at temperatures up to 1 170 K a silicon oxide layer forms.

In IR spectra of plasma polymer films deposited from benzene, there is an intensity decrease of the δ C–H deformation oscillation at 1 030 cm^{-1} with increasing power densities. This can be explained by the strong decrease in the proportion of unsubstituted benzene rings with increasing power density. The C–C stretching oscillation at 1 600 cm^{-1} barely changes in intensity with changing deposition power density. Other oscillations such as the ν C–C stretching oscillations at 1 450 cm^{-1} and 1 480 cm^{-1}, or the δ C–H deformation oscillation from the ring plane at 760 cm^{-1}, were also observed for all power densities. But all oscillations related to hydrogen decrease in intensity at higher power densities. This can be interpreted as a decrease in the hydrogen proportion.

Other authors were unable to further interpret the structure of plasma polymer films made from aromatic compounds [125]. A model structure of ethylene plasma polymer films has been suggested by [126] on the basis of IR spectra. This model proceeds on the assumption of many branched carbon chains, with bonds of aromatic rings (on a non-aromatic monomer!) and open bonds.

In principle, IR spectra confirm the polymer character of plasma polymer films. Characterizations of the plasma polymer according to the grade of polymerization, for example, or the chain length, are not very satisfying. The reasons lie in this special deposition technology. Only qualitative statements are possible, and even then with some reservations, about the degree of cross-linking.

With the assistance of photoelectron spectroscopy (XPS), the chemical binding relations for both the film surface and its interior can be determined by recording a depth profile. The chemical structure of plasma polymer films made from fluorocarbons and from silicon organic monomers has been studied extensively (e.g. [29,122]). For investigations of films made from hydrocarbons the method fails because there is no chemical shift from C–H bonds with the C 1s binding energy of 284.5 eV. But for investigations of the vertical film structure of films with embedded nanoparticles it is useful to know the etching rates for films made at various power densities. For that purpose, the layers can be argon-sputter-etched in steps until the substrate is reached. Since all films are deposited on thermally oxidized silicon wafers, the sudden onset of oxygen content indicates that the silicon oxide substrate layer has been reached. Argon etching rates were evaluated from the quotient of the film thickness d and the sputtering time t_{sp}.

The left-hand diagram of Fig. 2.3 shows the argon sputtering time t_{sp} (in seconds) needed to etch the complete film, as a function of plasma polymerization time for benzene plasma polymer films deposited with power densities of $p = 0.6$ W cm^{-2}. It can be seen that there is an approximately linear relationship between sputtering time and plasma polymerization time.

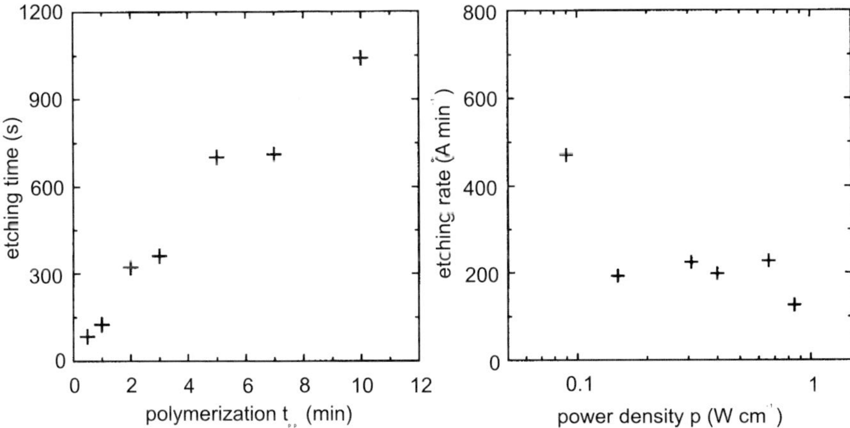

Fig. 2.3. *Left*: argon etching time t_{sp} in seconds for a complete off-sputtering of plasma polymer films made from benzene at $p = 0.6$ W cm^{-2}, as a function of polymerization time t_{pp} in minutes. *Right*: argon etching rate in Å min^{-1} as a function of power density p

The right-hand diagram of Fig. 2.3 shows the argon etching rate in Å min^{-1} as a function of the power density p. The etching rate decreases from 470 Å min^{-1} to 120 Å min^{-1} with increasing power density. From these etching rates, the loss of material can be determined for defined sputtering times. To compare these values, etching rates are given for carbon at 50 Å min^{-1} and for polymethylmethacrylate (PMMA) at 500 Å min^{-1} with argon sputtering 500 V and vertical incidence [127]. Values determined for benzene are transferable to plasma polymer films made from styrene.

Since detection of hydrogen is not possible with XPS, the ratio of carbon to oxygen is the only statement that can be made about the chemical structure of plasma polymer films. The oxygen content on the surface is 10–20%. No significant correlation with deposition conditions is noticeable. In fact, the oxygen proportion on the surface is mainly determined by the time of storage in UHV conditions before XPS analysis. It is assumed [128, 265] that the layer adsorbs the impinging O_2 molecules, but only a small amount is chemisorbed. Already after short sputtering times ($t_{\mathrm{sp}} \approx 15$ s, about 5 nm material removal) the oxygen content decreases to below 1%, and consequently beneath the detection limit. Comparing with IR spectra, it is reasonable to assume that oxygen-free plasma polymer films can be made using benzene and styrene. The formation of special monolayers at the substrate/plasma polymer film interface (e.g., NbC was found with plasma polymer films on Nb substrates [129]) has not been observed with XPS depth profile analysis.

Furthermore, no measurable shifts or splitting of the carbon peaks could be determined on plasma polymer films made from HMDSO. The formation of silicon carbide can be excluded. The elementary composition defined by XPS spectra on plasma polymer films made from HMDSO at $p = 0.85$ W cm^{-2} amounts to 70% C, 15% Si, and 15% O on the surface, and after 5 nm surface removal 65% C, 15% Si, and 17% O. Inside the layer, no further changes in elemental composition were measured. The measured elemental compositions are comparable to those from other investigations with AES or XPS (e.g., 68% C, 16% Si, 16% O in [92], 60% C, 25% Si, 15% O in [120], and 57% C, 25% Si, 16% O in [119]). From elemental analysis (EA), the composition of plasma polymer films made from HMDSO has been investigated and determined as 33% C, 45% Si, 13% O, 9% H [130] and 37% C, 40% Si, 15% O, 8% H [119]. The elemental composition of the monomer is 44% C, 35% Si, 10% O, 11% H [130]. The oxygen proportion grows with increasing power density [49]. The carbon content of HMDSN plasma polymer films decreases with higher substrate temperature, and at a substrate temperature above 900 K, carbon is no longer detectable [51]. With a further increase in temperature (1200 K), the plasma polymerization of HMDSN at similar power densities creates polycrystalline silicon carbide [131].

Film Surface. Since plasma polymer films form closed layers even at low thicknesses, it can be assumed that, with thin layers ($d \leq 50$ nm), the surface of a plasma polymer film conforms to the substrate surface. With higher film thicknesses, a special surface morphology can grow, known as the cauliflower morphology. In particular, for the deposition of plasma polymer with embedded metal nanoparticles (Sect. 2.2.2), the surface morphology of plasma polymer films is responsible for the growth process of the metal particles, and therefore for their particle size and shape distributions. Scanning electron microscopy (SEM) and atomic force microscopy (AFM) was therefore carried out to characterize the plasma polymer films surface. Whereas direct imaging is possible with AFM, investigations with SEM are much easier using a conductive film surface to avoid electrical charging of the film by the electron beam. This can be achieved by means of a 3 nm thin sputtered platinum film, for example.

Figure 2.4 shows four AFM micrographs of plasma polymer films made from benzene at different plasma polymerization power densities. The gray gradient (dark to white) describes a vertical height scale with maximum 60 nm. Direct imaging of the surface clearly shows that the cauliflower morphology grows with the power density of the plasma polymerization. At a power density of $p = 0.03$ W cm^{-2} there is no surface structure on the image,

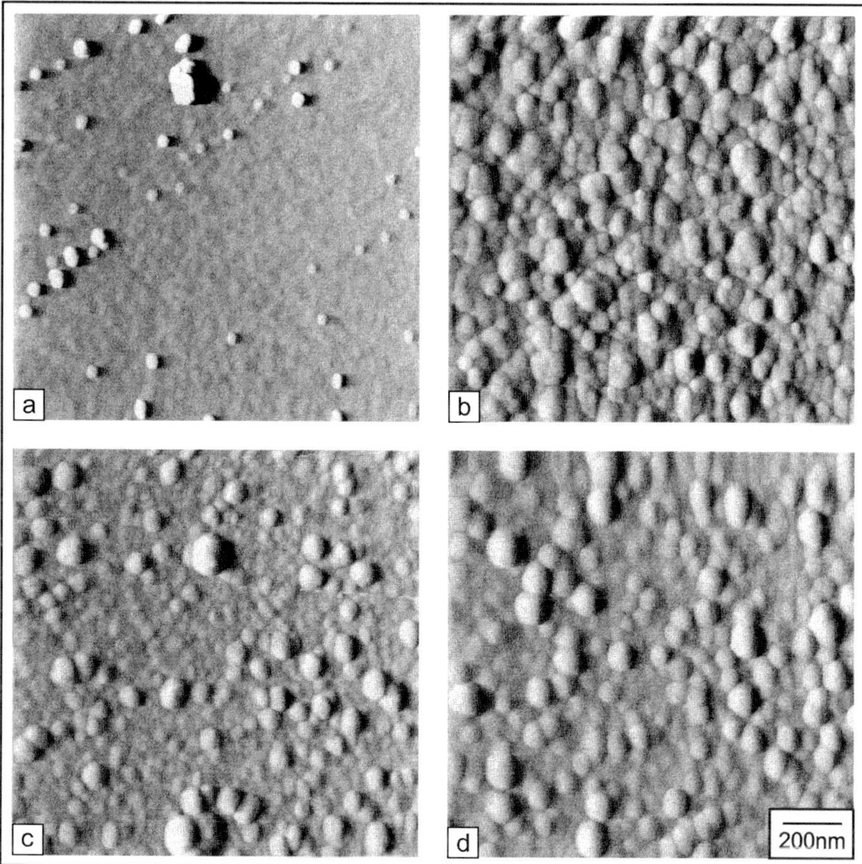

Fig. 2.4. Atomic force micrographs (AFM) of plasma polymer films made from benzene with different power densities: (*a*) $p = 0.03$ W cm^{-2}, $d = 50$ nm, (*b*) $p = 0.15$ W cm^{-2}, $d = 150$ nm, (*c*) $p = 0.40$ W cm^{-2}, $d = 180$ nm, (*d*) $p = 0.85$ W cm^{-2}, $d = 120$ nm

only dust particles. The cauliflower morphology starts to form at power densities of $p \geq 0.15$ W cm^{-2}. SEM investigations showed similar surface structures to Fig. 2.4, whilst on thicker films, larger structures could be found.

The cauliflower morphology has also been identified using AFM [132] and SEM investigations on various plasma polymer films made from polymethylmethacrylate [43, 132–138]. It is assumed that this morphology also develops by tensile stress in the film during deposition and further changes during aging. A model for the surface roughness of plasma polymer films has been given in [139].

The free surface energy or surface tension $\gamma_s = \gamma_{sd} + \gamma_{sp}$ can be determined by contact angle measurements and using the method for a three-medium air/liquid/solid system. Measurements are influenced by swelling and aging of the plasma polymer films. For plasma polymerized benzene films the values for the dispersive component γ_{sd} and polar component are γ_{sp} are $\gamma_{sd} = 30\text{–}35$ mN m^{-1} and $\gamma_{sp} = 5\text{–}10$ mN m^{-1}. This gives a total surface tension of $\gamma_{s} = 35\text{–}45$ mN m^{-1}. Other values were determined in [114] with $\gamma_{sd} = 25\text{–}35$ mN m^{-1} and $\gamma_{sp} = 0.3\text{–}2.3$ mN m^{-1}, where the higher values represent higher polymerization powers. Similar values are also found in [123], where the aging process of the films decreases the dispersive component of the free surface energy and increases the polar component. For films made from HMDSN or HMDSO, surface tension values of $\gamma_{sd} = 18\text{–}28$ mN m^{-1} and $\gamma_{sp} = 2\text{–}10$ mN m^{-1} were determined for various power densities. This corresponds to values of $\gamma_{sd} = 28$ mN m^{-1} and $\gamma_{sp} = 4.6$ mN m^{-1} given in [44]. For plasma polymer films made from HMDSO, the total surface tension increases with additional oxygen supply, from $\gamma_{s} = 19\text{–}40$ mN m^{-1} up to $\gamma_{s} = 72$ mN m^{-1} [140].

To sum up, for the various monomers and plasma polymerization properties, the surface tension of plasma polymer films varies widely and can be adjusted to meet specific requirements.

Optical Properties. As a preparation for discussing the optical properties of plasma polymer films with embedded metal nanoparticles in Sect. 6.1.2, we now describe optical properties of plasma polymer films without embedded metal in the UV–visible–NIR region. Optical transmission was measured in a wave number range from $4\,000$ cm$^{-1} \leq \tilde{\nu} \leq 50\,000$ cm^{-1} for films deposited on quartz substrates, using a conventional optical spectrometer. Optical reflection measurements were carried out using an Ulbricht sphere and different standards of reflection units between $4\,000$ cm$^{-1} \leq \tilde{\nu} \leq 40\,000$ cm^{-1}.

Figure 2.5 shows the UV–visible–NIR spectra of plasma polymer films made from benzene for different polymerization times t_{pp} with constant power density $p = 0.85$ W cm^{-2} (left) and for different power densities p with constant polymerization time $t_{pp} = 2$ min (right). All films show a transmittance which begins in the visual spectral range with high values and decreases rapidly into the UV region. This explains why films deposited at higher power density have a yellow color tint and thicker films look increasingly browner. Furthermore, as seen in Fig. 2.5 (left) the transmission spectra (d) and (e) are affected by interference. Decreased optical transmission of films deposited at higher power densities is caused by the slightly increased film thickness as much as by the changing chemical structure of the films.

In the following section, transmission spectra will be discussed in terms of molecular absorption. The excitation of $\sigma \to \sigma^*$ transitions is found at $\tilde{\nu} \approx 83\,000$ cm^{-1} [141, 142]. The longest wavelength (but least intensive) transition $(S_0 \to S_2)$ of benzene leads to absorption at $38\,000 \leq \tilde{\nu} \leq 43\,000$ cm^{-1}. The C=C double bond is excited by the $\pi \to \pi^*$ transition at $\tilde{\nu} \approx 55\,000$ cm^{-1}. However, the creation of polyacenes or n-conjugated –C=C–C= leads to a red shift. Even for naphthalene or a conjugation number of $n = 1$ [141], the absorption shifts into the investigated spectral region (see Fig. 2.5). The

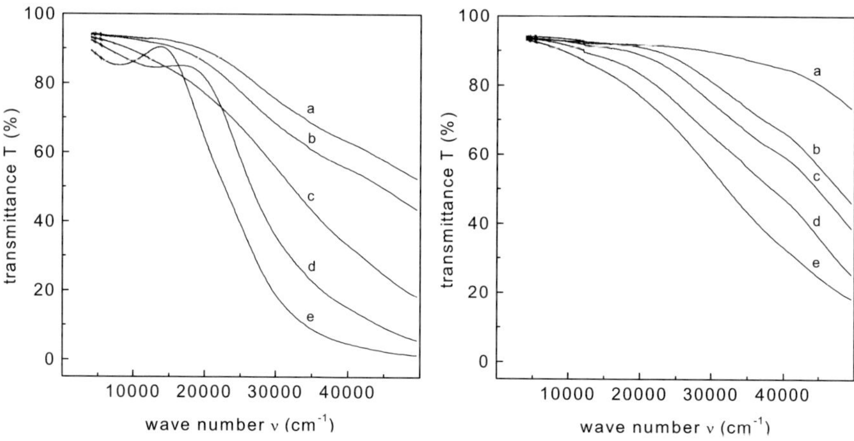

Fig. 2.5. UV–visible–NIR transmission spectra of plasma polymer films made from benzene. *Left*: constant power density ($p = 0.85$ W cm^{-2}) and different plasma polymerization times t_{pp}. (*a*) $t_{\mathrm{pp}} = 0.5$ min, $d = 40$ nm, (*b*) $t_{\mathrm{pp}} = 1$ min, $d = 50$ nm, (*c*) $t_{\mathrm{pp}} = 2$ min, $d = 120$ nm, (*d*) $t_{\mathrm{pp}} = 5$ min, $d = 260$ nm, (*e*) $t_{\mathrm{pp}} = 10$ min, $d = 350$ nm. *Right*: constant plasma polymerization time $t_{\mathrm{pp}} = 2$ min and different power densities p. (*a*) $p = 0.03$ W cm^{-2}, $d = 20$ nm, (*b*) $p = 0.09$ W cm^{-2}, $d = 50$ nm, (*c*) $p = 0.15$ W cm^{-2}, $d = 60$ nm, (*d*) $p = 0.40$ W cm^{-2}, $d = 90$ nm, (*e*) $p = 0.85$ W cm^{-2}, $d = 120$ nm

formation of conjugated double bonds is quite possible for plasma polymer films made from ring-formed hydrocarbons [126, 143]. If oxygen exists in the layer, the $\pi \rightarrow \pi^*$ transition of the carbonyl group C=O at $\tilde{\nu} \approx 34\,000$ cm^{-1} also contributes to optical absorption. With electron energy loss spectroscopy (EELS), excitations of π electrons at $\tilde{\nu} \approx 54\,000$ cm^{-1} were determined for a-C:H layers made from benzene. The excitation for graphite is at $\tilde{\nu} \approx 40\,000$–$50\,000$ cm^{-1} [125]. Due to nonuniform chain lengths and crosslinks between macromolecules, the transmission spectrum is characterised by a wide spectral distribution of single absorptions, and continuously decreasing transmission, especially in the UV region, with increasing wave number. Since no single excitation can be observed in the absorption spectra, few conclusions can be drawn concerning the chemical structure of the plasma polymer.

Figure 2.6 shows the UV–visible–NIR transmission spectra for five plasma polymer films made from HMDSN at different power densities before (left) and after (right) thermal treatment up to 495 K. Only small differences were found between the spectra before and after thermal treatment. The transmittance at $\nu = 34\,000$ cm^{-1} changes only from 0.18 to 0.22 for a film deposited with the lowest power density $p = 0.15$ W cm^{-2} (spectrum a). The spectra of films deposited at $p = 0.4$ W cm^{-2} (spectrum b), $p = 1.0$ W cm^{-2} (spectrum c) and $p = 1.8$ W cm^{-2} (spectrum d) are nearly identical before and after thermal treatment. For the film deposited with the highest power density

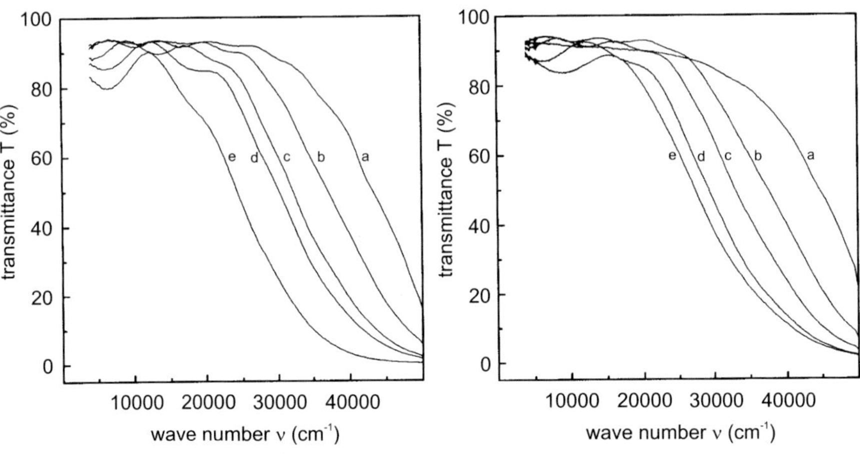

Fig. 2.6. Optical transmission spectra of plasma polymer films (HMDSN) before (*left*) and after (*right*) thermal treatment up to 495 K, for constant plasma polymerization time $t_{\mathrm{pp}} = 10$ min and different power densities p. (*a*) $p = 0.15$ W cm^{-2}, (*b*) $p = 0.4$ W cm^{-2}, (*c*) $p = 1.00$ W cm^{-2}, (*d*) $p = 1.8$ W cm^{-2}, and (*e*) $p = 3.00$ W cm^{-2}

$p = 3.0$ W cm^{-2}, the transmission in the whole spectral range was increased by about 10%, and interference effects shifted slightly.

Plasma polymer films made from benzene show similar behavior. After thermal treatment up to 480 K there are again no major changes in the transmission behavior. Benzene plasma polymer films therefore also possess good thermal stability up to this temperature.

The non-existent or small changes in transmission spectra of plasma polymer films suggest only very small changes in film thickness or chemical structure, and therefore good thermal stability at temperatures below 500 K.

Since optical transmission is very sensitive to the different polymerization power densities, reproducibility of fabrication conditions can be controlled by transmission spectra. Furthermore, it is possible to use transmission spectra to make statements about changes in film thickness, for example, after thermal treatment at higher temperatures, or after laser treatment.

Anticipating later descriptions of the optical properties of plasma polymer films with embedded metal nanoparticles, it is useful to calculate the film-thickness-independent complex dielectric function $\hat{\varepsilon}(\tilde{\nu}) = \varepsilon'(\tilde{\nu}) + \mathrm{i}\varepsilon''(\tilde{\nu})$ from the film-thickness-dependent transmission and reflection spectra. The index of refraction $n(\tilde{\nu})$ and the index of absorption $k(\tilde{\nu})$ can be determined from the dielectric functions using $\varepsilon' = n^2 - k^2$ and $\varepsilon'' = 2nk$.

Figure 2.7 shows the real part $\varepsilon'(\tilde{\nu})_{\mathrm{Is}}$ and the imaginary part $\varepsilon''(\tilde{\nu})_{\mathrm{Is}}$ of the dielectric function, the index of refraction n, and the optical band gap E_{04} for five plasma polymer films made from benzene at constant polymerization time $t_{\mathrm{pp}} = 3$ min with power densities of $p = 0.15$ W cm^{-2} (a), $p = 0.4$ W cm^{-2}

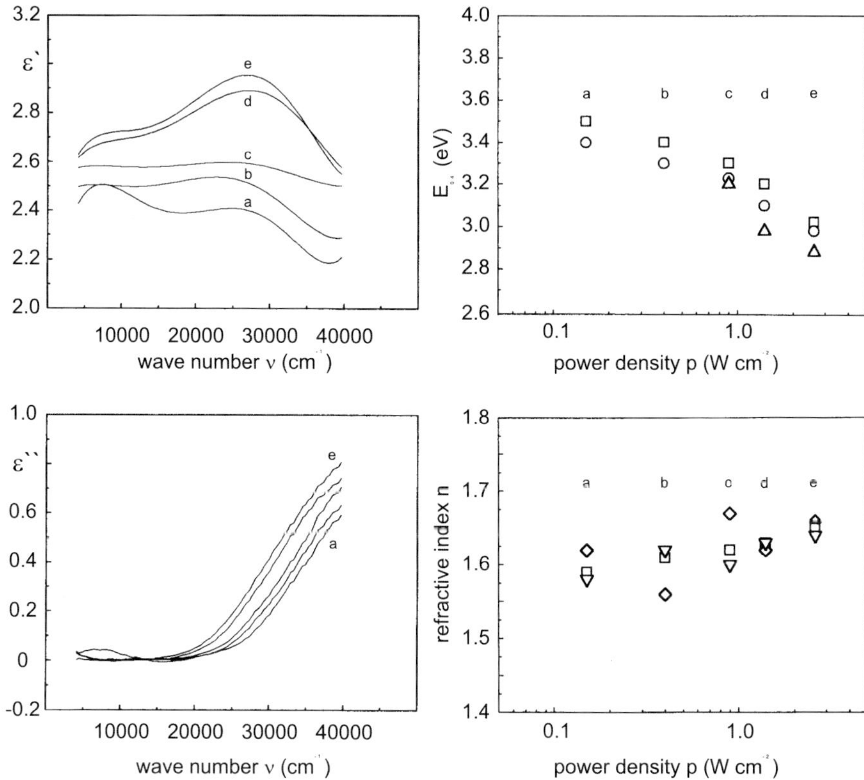

Fig. 2.7. Real part ε' (*top left*) and imaginary part ε'' (*bottom left*) of the dielectric function $\varepsilon(\nu)$, together with E_{04} (*top right*) and the index of refraction n (*bottom right*) for five plasma polymer films made from benzene with constant polymerization time $t_{\mathrm{pp}} = 3$ min and different power densities p (after [146])

(b), $p = 1.0$ W cm^{-2} (c), $p = 1.8$ W cm^{-2} (d) and $p = 3.0$ W cm^{-2} (e). It can be seen from the figure that the dielectric functions with the smallest real and imaginary parts belong to the film deposited at lowest power density. The values for the real part $\varepsilon'_{\mathrm{Is}}$ and $\varepsilon''_{\mathrm{Is}}$ increase little with higher power density.

The values for the index of refraction n were determined to lie in the range $1.55 \leq n \leq 1.65$ for the measured wave number range. In fact, n increases with rising power density. This dependence of n on the power density makes it possible to generate a vertical gradient in the refractive index by changing the polymerization power density. This may have applications in the development of optical light wave conductors, for example.

Refractive indices $1.5 \leq n \leq 1.9$ [118], $1.45 \leq n \leq 1.55$ [144] or $1.5 \leq n \leq 1.7$ [114] $1.45 \leq n \leq 1.55$ have been determined for plasma polymer films made from HMDSO, whereas higher values for higher power densities have been observed in [114, 144]. HMDSN plasma polymer films generally have an index of refraction $n \approx 1.5$ [91, 92, 119], but refractive indices in the range $1.9 \leq n \leq 2.1$ have also been determined [35].

The optical energy E_{04} is defined as the energy for which the absorption coefficient $\alpha = 10^4$ cm^{-1}. If $(\alpha E)^{\frac{1}{2}}$ depends linearly on E, then the optical band gap E_{opt} can also be determined from $(\alpha E)^{\frac{1}{2}} = G(E - E_{\mathrm{opt}})$ [19, 145]. The proportionality factor G is assumed to be constant. However, this linear dependence does not usually hold for plasma polymer films made from benzene.

Values for E_{04} decrease with increasing power density, and lie between $E_{04} = 3.4 \pm 0.1$ eV for $p = 0.15$ W cm^{-2} and $E_{04} = 2.9 \pm 0.2$ eV for $p = 2.5$ W cm^{-2} (Fig. 2.7). For plasma polymer films made from different monomer mixtures such as C_2F_6 and C_2H_2, values of E_{04} lie in the range $E_{04} = 2.5$–2.8 eV [147]. For HMDSO plasma polymer films the optical band gap was determined to be $E_{\mathrm{opt}} = 4.7$–5.2 eV [144]. The values for plasma polymer films made from a monomer mixture of chlorotrifluoroethylene (C_2F_3Cl) and benzene have been determined as $E_{\mathrm{opt}} = 3.4$ eV (monomer 100% C_2F_3Cl) and $E_{\mathrm{opt}} = 1.6$–2.2 eV (monomer mixture 60–15% C_2F_3Cl) [148]. For a-C:H layers made from benzene, the values are $E_{04} = 1.0$–2.0 eV and $E_{\mathrm{opt}} = 0.8$–1.8 eV [115, 149], well below the values of Fig. 2.7. However, they increase with increasing hydrogen proportion [115]. The higher values for E_{04} compared with a-C:H layers explain the polymer character of plasma polymer films made from benzene.

Electrical Properties. To determine the electrical d.c. conductivity, the plasma polymer films were deposited on a conductive substrate (silicon wafer with deposited gold electrodes). After the plasma polymer deposition, a gold top layer was deposited on the film as electrode. The film was then contacted with micro-tips or conductive silver lacquer to avoid mechanical stress introduced by the microelectrodes. The resulting sandwich structure was used to measure the current–voltage characteristic (I–U relationship) at applied electrical field strengths $E = U/d$ in the range 10^2 V cm$^{-1} < E(\sigma) < 10^6$ V cm^{-1}, where U is the applied voltage and d the film thickness. These measurements are the basis for determining the electrical d.c. conductivity $\sigma(E)$ using the sample geometry.

Figure 2.8 shows the current–voltage characteristics for several plasma polymer films made from HMDSN. The left-hand diagram presents the I–U characteristic for four films made at a power density of $p = 0.14$ W cm^{-2} with different film thicknesses produced using different plasma polymerization times t_{pp}. The graphs are almost linear until the breakdown voltage is reached. The right-hand diagram shows the I–U characteristics for four plasma polymer films made with different power densities, but with the same plasma polymerization time $t_{\mathrm{pp}} = 5$ min. In fact, the current–voltage characteristic is nearly linear for $E \leq 10^5$ V cm^{-1}, and conductivities increase with increasing deposition power densities. The conductivity is significantly higher for films deposited at the highest power density $p = 1.88$ W cm^{-2}. Note, how-

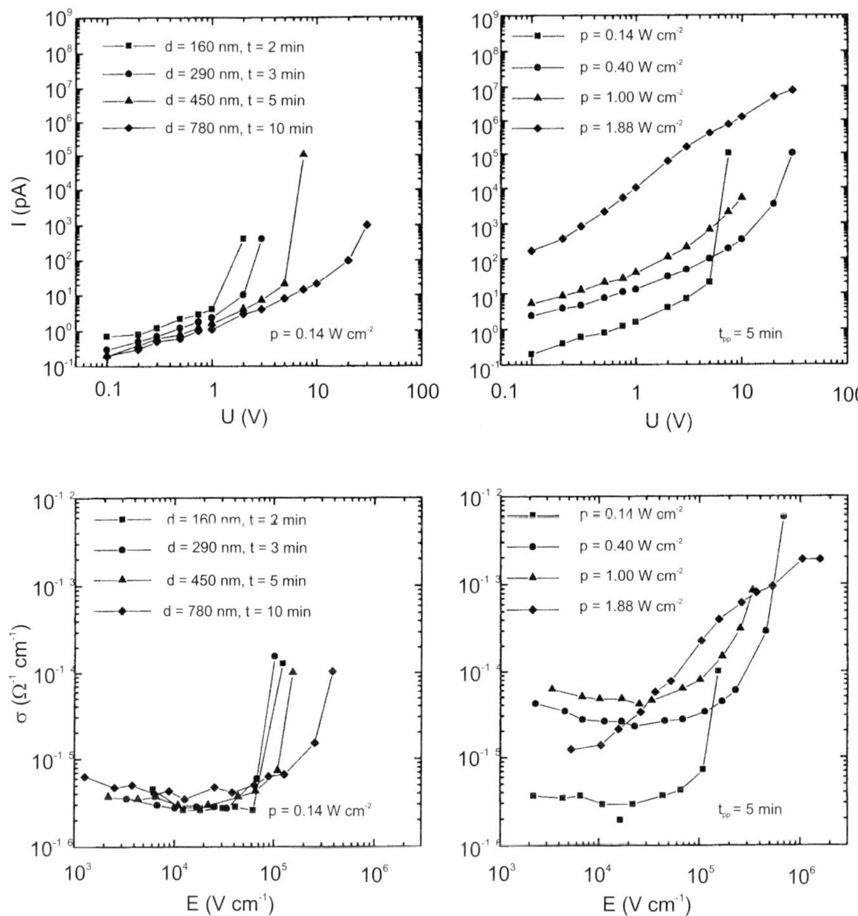

Fig. 2.8. Current voltage characteristic (*top*) and electrical d.c. conductivity $\sigma(E)$ (*bottom*) for plasma polymer films made from HMDSN for different film thicknesses (*left*) and different power densities (*right*)

ever, that the different film thicknesses of the four layers ($p = 0.14$ W cm^{-2}, $d = 450$ nm; $p = 0.40$ W cm^{-2}, $d = 440$ nm; $p = 1.00$ W cm^{-2}, $d = 280$ nm; $p = 1.88$ W cm^{-2}, $d = 200$ nm) complicate direct comparisons of I–U characteristics.

The field strength dependent d.c. conductivity $\sigma(E)$ can be calculated from the I–U relationship and the formula

$$\sigma(E) = \frac{Id}{A_{\mathrm{el}}U} \quad \text{and} \quad E = \frac{U}{d} \, ,$$

where d is the film thickness, I the current, A_{el} the electrode area, and U the voltage. This is also shown in Fig. 2.8. For films deposited with a power density of $p = 0.14$ W cm^{-2}, the d.c. conductivity reaches values of $\sigma(E) = 2\text{--}4 \times 10^{-15}\,\Omega^{-1}$cm^{-1}, independently of the film thickness, at field strengths $E \leq 10^5$ V cm^{-1}. The breakdown voltage of $E \approx 1\text{--}5 \times 10^5$ V cm^{-1} depends on the film thickness. Films made at higher power densities $p \geq 0.4$ W cm^{-2} show conductivities of $\sigma(E) = 6\text{--}8 \times 10^{-14}\,\Omega^{-1}$cm^{-1}, which is about one order of magnitude higher. For films made at power densities $p < 1.0$ W cm^{-2}, the d.c. conductivity is nearly constant up to an electrical field strength of $E = 10^5$ V cm^{-1}. The slightly higher values at field strengths $E < 10^4$ V cm^{-1} result from the electrostatic charge of the plasma polymer films which influences measurements. The increase in d.c. conductivity at field strengths $E > 10^5$ V cm^{-1} can be caused by the onset of electrical breakdown. As shown in the I–U characteristics of Fig. 2.8, the film made at the highest power density $p = 1.88$ W cm^{-2} exhibits a completely different conductivity behavior. The conductivity $\sigma(E)$ increases almost constantly with electrical field strength, and there is no observed breakdown voltage.

The experimental results for the electrical d.c. conductivity of plasma polymer films are comparable to those reported by other authors. The d.c. conductivities of plasma polymer films made from styrene lie in the range $\sigma = 10^{-15}\text{--}10^{-16}\,\Omega^{-1}$cm^{-1} [150] and for HMDSO films between $\sigma = 10^{-15}\,\Omega^{-1}$cm^{-1} [151] and $\sigma = 2 \times 10^{-16}\,\Omega^{-1}$cm^{-1} [102]. For plasma polymer films made from styrene, the breakdown field strengths ranged over 1×10^4 V cm$^{-1} \leq E \leq 4 \times 10^6$ V cm^{-1} [152] depending on the polymerization power density. The breakdown field strength lies in the range 8×10^5 V cm$^{-1} \leq E \leq 5 \times 10^6$ V cm^{-1} [153] for HMDSO films. Breakdown field strengths of about 8×10^7 V cm^{-1} were determined for plasma polymer films made from ethyle and trifluoromethane in [154].

Metal-free plasma polymer films made from conductive or iodine-doped monomers exhibit the following maximal conductivities: thiophene $\sigma = 1.3 \times 10^{-6}\,\Omega^{-1}$cm^{-1} [40], thiophene doped with iodine $\sigma = 10^{-4}\,\Omega^{-1}$cm^{-1} [38] and $\sigma = 10^{-4}\text{--}10^{-3}\,\Omega^{-1}$cm^{-1} [155, 156], acrylnitrile doped with iodine $\sigma = 10^{-5}\,\Omega^{-1}$cm^{-1} [157], perylene with different substitutes $\sigma = 10^{-1}\Omega^{-1}$cm^{-1} [158], tetracyanoquinodimethane (TCNQ) $\sigma = 10^{-6}\Omega^{-1}$cm^{-1} [159], TCNQ and quinoline $\sigma = 10^{-5}\Omega^{-1}$cm^{-1} [160], and 2-iodo-thiophene $\sigma = 10^{-1}\Omega^{-1}$cm^{-1} [39]. Some plasma polymer films, e.g., from phenylacetylene [161] or styrene [162] show an acceptable, spectrally dependent photoconductivity.

To describe electrical transport in thin plasma polymer films, electrons and holes are considered as charge carriers. Ionic conductivity can usually be neglected, although specific fabrication of ionically conducting plasma polymers is possible [163].

The injection of charge carriers into plasma polymer films takes place by Schottky emission from the metal electrode or by Poole–Frenkel generation of electron–hole pairs. If these charge carrier generation processes are slower than the charge carrier transport, they control the electrical behavior of the sample. When charge carrier transport is slower than generation, electrical transport properties are explained by space charge limited currents (SCLC theory). For plasma polymer films made from benzene and styrene, and also for HMDSN and HMDSO, it is assumed that Poole–Frenkel generation is the dominating process for d.c. conductivity, in conjunction with space charge limited currents [164, 165].

Results of measurements depend strongly on measurement methods and the electrodes used. Changes in d.c. conductivity have also been observed through the influence of oxygen and moisture. In [166], multiple repeatable switching was noticed in plasma polymer sandwich structures (film thickness 70–150 nm) with silver electrodes at field strengths of $E \approx 2 \times 10^6$ V cm^{-1}. In this case and also in [167] the effect is explained by the formation of a metallic conducting path through the plasma polymer films. More detailed surveys of electrical conductivity in plasma polymer films are given in e.g. in [165].

The results presented show that the electrical d.c. conductivity of plasma polymer films can be varied by an appropriate choice of fabrication conditions (plasma polymerization power density), and that the matrix conductivity for embedded metal nanoparticles can thereby be influenced.

2.2 Deposition of Polymer Thin Films with Embedded Metal Nanoparticles

2.2.1 Embedding Metal Nanoparticles in Polymer Films

Bearing in mind the general focus on vacuum technologies for depositing polymer films with embedded metal nanoparticles, as mentioned in Chap. 1, only these technologies will be described here. This includes technologies pertaining to chemical vapor deposition (CVD) or physical vapor deposition (PVD). Besides vacuum technologies for the production of metal-containing polymer thin films, there are a large number of non-vacuum fabrication methods.

Polymer films with embedded metal nanoparticles can be made with metal particles that are generated by growing them from a solution and stabilizing them with shells of ligand molecules or more simply with gelatine. Otherwise a metal colloid solution can be added to a monomer, so that metal-containing films are generated by self-organization. These deposition methods and properties of the deposited films are reviewed in [6, 15, 18, 168].

Polymer films with embedded metal nanoparticles can also be fabricated by simply melting a powder mixture (Co alloy in polyethylene [169]), by drying in a powder resin mixture (Ag particles in epoxy resin [170]) or by moulding a metal–polymer mixture (Ag particles in PTFE [171]). In addition, there are deposition methods on the borderline between vacuum and non-vacuum technologies. Metal-containing polymer films can also be produced, if metal particles generated by evaporation in an inert gas atmosphere are put into a dissolved polymer or into a polymerizable solution (Ag particles in polyvinylalcohol solution [172] or Au particles in propanol [173]).

The most common vacuum technologies are thermal evaporation, non-reactive or reactive sputtering, and plasma polymerization. With these techniques, solid films grow from the gas phase. In principle, deposition processes can be divided into non-competitive and competitive growth of polymer and metal. In non-competitive growth, polymer films and metal particles grow independently from the other material at the same time or consecutively. In competitive growth, both materials grow at the same time and the growth behaviour of one material is influenced to some extent by the growth of the other material.

The following deposition methods for polymer films with embedded metal particles are mainly characterized by non-competitive growth:

- alternating sputtering,
- alternating evaporation,
- alternating plasma polymerization and metal evaporation.

Deposition methods involving competitive growth are mainly:

- co-sputtering,
- co-evaporation,
- simultaneous plasma polymerisation and metal sputtering,
- simultaneous plasma polymerisation and metal evaporation,
- organometallic CVD (OM-CVD).

In the non-competitive growth mode, metal particle growth is similar to that observed in thin, discontinuous metal films. In most cases, the binding energy of the deposited metal is higher than the binding energy between metal and substrate, an island growth (Volmer–Weber growth) results and metal nanoparticles form. A continuous layer forms when the islands are large enough to begin to grow together. Other growth modes, such as Frank and van der Merve's monolayer growth, or the conversion of an initial monolayer growth mechanism into island growth after Stranski and Krastanov, seldom play an important part in metal particle growth.

If there is a similar nanostructure of thin, discontinuous metal films and metal nanoparticles embedded in polymer films deposited by non-competitive growth, embedding into the polymer results in effective passivation of the nanoparticles. Thin, discontinuous metal films are often only stable under vacuum conditions, becoming unstable in air. Particle diffusion is possible on the substrate because of a high substrate surface mobility and leads to a modified particle size and shape distribution (e.g., gold particles [174]). Furthermore, adsorption of molecules on the particle surface can lead to a change in electrical conductivity (e.g., Pd particles [175]). The high surface energy of the metal particle may cause particles of usually stable metals to oxidize, and other reactions may take place. For example, silver sulfide formation has been detected [176]. Passivation of discontinuous metallic thin films can be accomplished by deposition of a dielectric coating onto the metal particles. This can be made from ceramics or polymers.

Polymer films with embedded nanoparticles can also be fabricated by so-called relaxative autodispersion. A metal layer is first evaporated onto a polymer film. Particles then enter the polymer by a subsequent thermal treatment. Ag, Au, Ge, Pd and Au/Pd or Au/Co bimetallic nanoparticles in polyamide films have been reported [177–184].

In the competitive growth mode, where simultaneous deposition of the polymer and metal takes place, growth of embedded metal particles is substantially influenced by the coincidental growth of the polymer. For example, in co-sputtering, the size and shape distribution of the embedded metal nanoparticles can be adjusted by altering sputter parameters. Sputter yields for polymers and metal can be very different. This has to be taken into account when choosing the composite target design and deposition parameters to prevent target poisoning, for example. Polymer films with embedded metal particles made by co-sputtering are Au–PTFE, [185–187], Ag–PTFE [187], Cu–PTFE [186, 188] or Pt PTFE [186]. In addition, carbon films with embedded nanoparticles such as Ag [189] have been prepared by co-sputtering. Unlike polymer–metal co-sputtering, the co-sputtering deposition of ceramic materials with embedded metal particles is very common, using two different targets or a composite target, e.g., Ag–SiO_2 [190–193], Au–Al_2O_3 [194], Ag–Al_2O_3 [196], Au–Al_2O_3 [195], Au–MgO [197–199], Au–MoS_2 [200], Au–SiO_2 [190, 191, 193, 195, 202], Ge–SiO_2 [201, 203, 204], Si–SiO_2 [205], Pt–SiO_2 [195], Pt–Al_2O_3 [206] and Ni–Al_2O_3 [195, 196, 207] as well as Co–BN [208], Fe–BN [209], Fe–SiO_2 [210], Fe–Al_2O_3 [211] or Fe/Ni-alloy particles in SiO_2 or Al_2O_3 [212].

A further possible deposition technique involving competitive growth of polymer and metal is simultaneous thermal evaporation of the metal and evaporation (or sublimation) of the polymer. This is called co-evaporation, and uses resistance or electron beam evaporators. The embedding of Cu nanoparticles in polyethylene [213] or Ag in poly(octamethylcyclotetrasiloxane) [214] have been reported.

Again, the deposition of such films started with ceramic hosts and materials Ag–SiO_2 [215], Au–Al_2O_3 [216], Au–TeO_2, [217], Bi–SiO_2 [218, 219], Cu–PbI_2 [220], Co–Al_2O_3 [216, 221] and Pt–Al_2O_3 [222].

The other deposition methods characterized by competitive growth, that is, simultaneous plasma polymerisation and metal sputtering or simultaneous plasma polymerisation and metal evaporation, are described in detail in Sect. 2.2.2.

2.2.2 Plasma Polymer with Embedded Metal Nanoparticles

Metal-Containing Plasma Polymer Films. Metal-containing plasma polymer thin films can be made using the following methods:

- plasma polymerization of organometallic monomers,
- plasma polymerization of a monomer with admixture of an organometallic compound,
- plasma polymerization and metal evaporation.

A distinction can be made between two forms of the embedded metal:

- the metal occurs in a chemically bound form,
- the metal occurs in the form of small particles.

The use of organometallic monomers for plasma polymerization [32, 223–225] usually yields chemically bound metal atoms incorporated into the polymer, e.g., plasma polymerization with tetramethylgermanium [226–228], but can also produce a form of metal or metal oxide particles, e.g., plasma polymerization with tetramethyltin yields Sn particles [229, 230], indium acetylacetonate yields In_2O_3 particles [86], allylcyclopentadienylpalladium yields Pd particles [231] and titaniumtetraisoprepoxide yields Ti particles [93, 232].

Furthermore, the metal atoms are mostly chemically bound into the monomer with an admixture of metallic or organometallic compounds, e.g., hydrocarbon compounds with an admixture of $Fe(CO)_5$ [225, 233, 234] or methylmethacrylate with tetramethyltin [235]. Plasma polymer films with high metal proportions are made from gas mixtures of propane with allylcyclopentadienylpalladium or propane with dimethyl-(2.4-pentaneionato)gold [236].

Two other methods involve combinations of plasma polymerization with a further vacuum technology. The technology of simultaneous plasma polymerization and metal sputtering was developed by Kay [29] and enables the embedding of sputterable metals, metal alloys and polymers in different plasma polymer matrix materials. On the other hand, thermal evaporation can be carried out during a plasma polymerization. With this method many different metals, and dyes adapted for vacuum sublimation (such as phthalocyanine and rhodamine [237]), can be successfully embedded in a polymer matrix.

Both technologies, plasma polymerization with metal sputtering and plasma polymerization with metal evaporation, have been applied to make polymer films with embedded metal nanoparticles for the investigations described in this book. The deposition method and the resulting films will be explained in detail in the next two sections. Experimental investigations of the gas phase processes and the growth behavior of the plasma polymer films and metal particles are not considered further, being outside the scope of this book.

Ever since the earliest work with plasma polymer films containing embedded metal particles, there have been patents for technical uses of these films, e.g., as electronic devices [238, 239] and as conductor materials [240]. Given their nanodispersive structure, specific modifications of the films are possible via laser irradiation and electron irradiation. For example, laser irradiation was carried out with the goal of evaporating the plasma polymer on the irradiated spot and making a highly conductive gold conductor [241, 242]. Apart from this, metallic nanowires have been generated using ultrashort laser pulses [243, 244]. The specific lateral nanostructural modification of plasma polymer films with embedded metal nanoparticles is discussed in Sect. 4.3 for the case of laser irradiation and in Sect. 4.4 for electron irradiation.

Applications of styrene plasma polymer films with embedded gold particles for electron beam lithography and as an absorber material for X-ray lithography mask fabrication have been investigated in [245]. Further investigations aim towards future applications in electroluminescence [246], use as electron beam resist [93], and field-induced changes in electrical d.c. conductivity [247–249]. Films with a vertical copper gradient have been investigated as intermediate layers to improve the adhesion of copper on different polymers [250, 251].

Plasma Polymerization and Metal Sputtering (PPMS). Figure 2.9 shows the setup for a high vacuum reactor designed to produce metal-containing plasma polymer films by simultaneous plasma polymerization and metal sputtering. This was developed by Kay [29, 252] and has since been used by others [253–255]. The most significant elements of the deposition system in this parallel plate reactor are the two horizontal metal electrodes for plasma polymerization. These are thermally controlled and the upper electrode also serves as the sputter target. Substrates are placed on the lower electrode. The radio frequency (13.56 MHz) is supplied to the target by a matching network (coil and capacitor). The power of the r.f. generator can be controlled up to 40 W. Reflected power is minimized with the help of the matching box before deposition begins. The power density for plasma polymerization ranges over $p_{rf} = 0.15$–5 W cm^{-1}. During the deposition process, a self-biasing effect (up to 1500 V) can occur on the lower electrode. The chemical structure of the plasma polymer can be influenced by an additional bias voltage on the lower electrode [29].

Before argon is introduced into the reactor, the vacuum chamber is evacuated (high vacuum), and a predefined argon flow rate is supplied (100 Pa l s^{-1}) with the help of a controllable valve. A monomer is then admixed with the argon, and the argon flow rate decreases, whereas the total flow rate stays constant, independently of the gas mixture. The metal content in the plasma polymer film is adjusted by the ratio of argon to monomer. With an increasing monomer proportion, the sputtering rate on the metal target decreases, and the deposition rate of the plasma polymer increases. Even with a monomer proportion of 25% of the total flow rate, no metal will be sputtered off the target, and a metal-free plasma polymer film is produced.

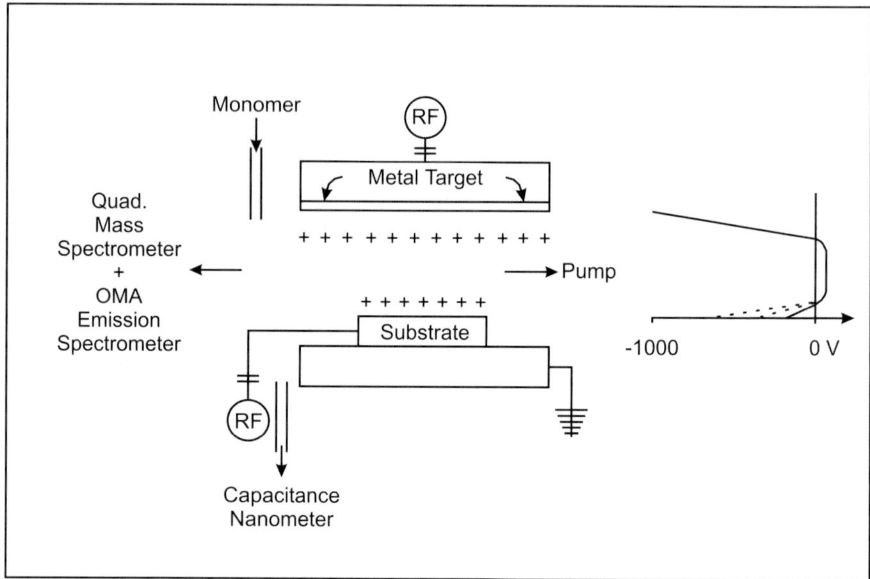

Fig. 2.9. Reactor for the fabrication of metal-containing plasma polymer films by simultaneous plasma polymerization and metal sputtering [256]

The method of simultaneous plasma polymerization and metal sputtering offers the possibility of producing films with a very homogeneous vertical distribution of embedded metal particles. The depth profiles determined using photoelectron spectroscopy (XPS) demonstrate this effect (see Sect. 3.2.2 and Fig. 3.15).

The following is a brief survey of monomers and metals used for film deposition by simultaneous plasma polymerization and metal sputtering. The most frequently investigated plasma polymer films with embedded particles are fluorine hydrocarbons with embedded gold particles, although recently, fluorine hydrocarbon polymers have been increasingly substituted by hydrocarbon polymers.

- perfluoropropane C_3F_8 with embedded Au, Mo, Al, Co, Ge or Cu nanoparticles [247, 252, 256–265],
- tetrafluoroethylene C_2F_4 with embedded Au or Mo nanoparticles [241, 242, 256, 261, 266, 267],
- hexafluoroethane C_2F_6 with embedded Au or Mo nanoparticles [37, 254, 262],
- octafluoro-cyclobutane C_4F_8 with embedded Au [268]
- tetrafluorocarbon CF_4 with embedded Au, Al, Co or Mo nanoparticles [262, 269–272],
- perfluorobutane C_4F_{10} with embedded Mo nanoparticles [262],
- chlorotrifluoroethylene C_2ClF_3 with embedded Au nanoparticles [271–278],

- methane CH_4 with embedded Cu nanoparticles [250],
- propane C_3H_8 with embedded Au, Co or Cu nanoparticles [246, 248, 253, 266, 279–285]
- n-hexane with embedded Ge, Mo or Ni [286–289].

Since the transition from plasma polymer films to a-C:H layers and hard films is gradual, a reactor of the kind shown in Fig. 2.9 can be used to produce metal-containing a-C:H layers. For example, Ti [255, 290, 291], W [291, 292] Ta [255, 293] or Au [294] have been embedded in a-C:H layers. The metals were sputtered in the presence of acetylene or other hydrocarbons. During sputtering, acetylene was added It is mainly the mechanical properties of these materials that incite interest [295, 296]. A similar deposition was also used for the plasma deposition of SiO_2 from SiH_4 together with r.f. sputtering of gold [297].

Plasma Polymerization and Metal Evaporation (PPME). In addition to the two plasma polymerization electrodes, an evaporation system is integrated into the reactor for the combination technology of plasma polymerization and metal evaporation. A thermal tungsten evaporator is supplied with high current at 40–100 A (voltage 12 V), and thus heated to evaporate the metal. The evaporator is placed directly beneath a recess in the lower electrode. Figure 2.10 shows the reactor setup, and the configuration of electrodes and evaporator. Analogously to the production of metal-free plasma polymer films in Sect. 2.1.3, a high voltage (500–2000 V, 50 Hz) is supplied to the electrodes to create the plasma.

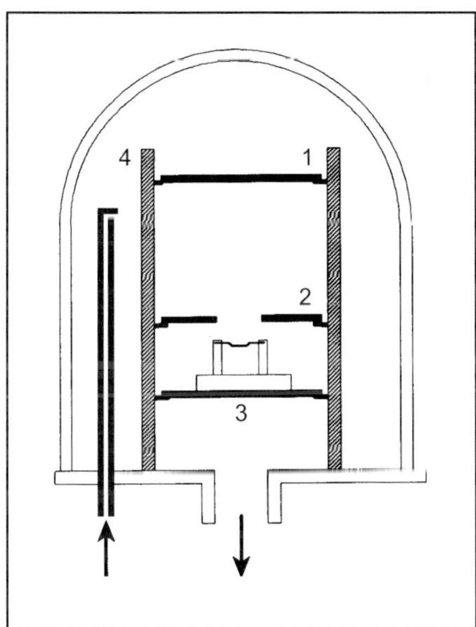

Fig. 2.10. Reactor setup for the fabrication of plasma polymer with embedded metal nanoparticles by simultaneous or alternating plasma polymerization and metal evaporation. (*1,2*) plasma electrodes, (*3*) evaporator, (*4*) PTFE supports

The spatial distribution of metal atoms depends on the evaporator temperature and monomer gas pressure, and is very different to the spatial distribution of metal atoms in high vacuum because of collision processes with monomer molecules and radicals. Using Monte Carlo simulations carried out for sputtering processes of alloys [298], for example, assumptions can be made about the spatial distribution of metal atoms in plasma discharges, and about the number of atoms arriving at the substrate as a function of the evaporator–substrate distance. An area of about 100 cm^2 of metal-containing film is deposited with approximately constant particle size distribution on the underside of the upper electrode (plasma polymer multilayer with nearly constant metal particle content). Films with embedded nanoparticles are also deposited on the upper side of the lower electrode as a result of collision processes between metal atoms and monomer molecules and fragments. The films show a lateral gradient in the metal content (plasma polymer multilayer with continuously varying metal particle content).

The fabrication of plasma polymer multilayer systems with embedded nanoparticles was carried out using the following procedure. Deposition starts with a plasma polymerization. After a certain time, usually 30–120 s, metal evaporation is started. This continues for 15–30 s and then plasma polymerization is continued for a further 30–120 seconds. If there is no interruption in the plasma polymerization whilst metal evaporation takes place, this fabrication method is referred to as simultaneous plasma polymerization and metal evaporation (SPPME). If plasma polymerization is stopped during metal evaporation, and reinitiated afterwards, the plasma polymer films with embedded nanoparticles are made by alternating plasma polymerization and metal evaporation (APPME). Since the growth rate of the metal is much higher than that of the plasma polymer, differences in metal embedding between the two methods (simultaneous and alternating plasma polymerization) are not very large, and metal particles are approximately embedded in one plane. The multilayer system produced using these methods, with film thickness d, consists of a first plasma polymer layer with thickness d_{p1}, a metal-containing layer with film thickness d_c, and a second plasma polymer layer with thickness d_{p2}. The separate film thickness of each layer can be determined by cross-section TEM investigations (see Sect. 3.2.2). It is possible to extend this multilayer system by further depositions of plasma polymer and metal-containing plasma polymer films (e.g., a multilayer system consisting of two or three metal nanoparticle planes), whereas a vertical metal gradient can be produced by evaporation of different metal amounts from several evaporators which are inserted in the deposition reactor near the first evaporator.

A nearly homogeneous vertical distribution of embedded metal particles can be achieved using several evaporators heated one after the other. The monomer flow rate is raised to increase the plasma polymer deposition rate, and the timing of deposition is controlled in such a way that metal evaporation starts simultaneously with plasma polymerization. Examples of the monomers used and of the embedded particles are:

- tetrafluorocarbon CF_4 with embedded Au nanoparticles [299, 300],
- butane C_4H_{10} with embedded Au, Ag, Cu or Pd nanoparticles [231, 301, 302],
- butyl C_4H_6 with embedded Ag or Te nanoparticles [303],
- benzene C_6H_6 with embedded Ag, Au, Sn or In nanoparticles [265, 304–324],
- benzene C_6H_6 with embedded Ge or Ho [309, 325],
- styrene $C_6H_5CH{=}CH_2$ with embedded Au [245], Au [265], In [319, 326] and Te [327] nanoparticles,
- chlorobenzene C_6H_5Cl with embedded Ag nanoparticles [311, 328],
- thiophene C_4H_4S with Ag nanoparticles
- hexamethyldisilazane $(CH_3)_3SiNHSi(CH_3)_3$ or hexamethyldisiloxane $(CH_3)_3SiOSi(CH_3)_3$ with embedded Ag [304, 310, 329–335] Cu [336, 337] or Sn [336] nanoparticles ,
- vinyltrimethylsilane $H_2C{=}CHSi(CH_3)_3$ with embedded Au [338, 339] or Ag [340] nanoparticles
- tetraethoxysilane with embedded Ag nanoparticles [338, 340]
- polyvinylcarbazole (sublimated into the plasma discharge) with embedded Au [341],
- triallylphosphine with Ni [225],
- carbon disulfide CS_2 with embedded Te or Bi [327].

With the conditions chosen for fabrication of metal-containing plasma polymer films (substrate at room temperature, absolute pressure inside the reactor $\approx$ 1 Pa), island growth of evaporated metals (gold, silver, copper) takes place. Although the size distribution of metal particles is similar to that in discontinuous metallic thin films, plasma polymer films with embedded metal nanoparticles have the advantage of long-term stability, because the metal particles are located in a matrix, and changes caused by their storage in air (e.g., gas adsorption, oxidation) are prevented or occur only very slowly. Furthermore, the kind of surface diffusion processes found with discontinuous metal films are prevented by the plasma polymer matrix.

The size of the embedded polycrystalline metal particles is determined by the distance from evaporator to substrate, the evaporated amount of metal and also the temperature of the evaporator. Furthermore, with simultaneous plasma polymerization and metal evaporation the growth behavior of the metal particle is weakly influenced by plasma polymerization. Since metal evaporation takes place later than plasma polymerization during the fabrication of multilayer systems, the surface of the basic polymer films greatly influences the particle distribution (Sect. 3.2.1).

Plasma polymer multilayers with embedded particles. The fabrication of multilayer systems by simultaneous plasma polymerization and metal

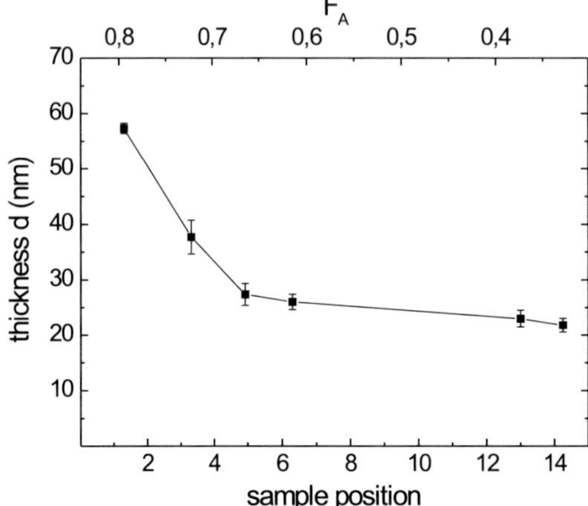

Fig. 2.11. Film thickness as a function of substrate position for plasma polymer multilayers made from HMDSN with embedded silver nanoparticles

evaporation ideally complements the technology described in Sect. 2.2.2 for the fabrication of metal-containing plasma polymer films by plasma polymerization and metal sputtering. Films made by simultaneous plasma polymerization and metal sputtering show a vertically and horizontally homogeneous particle distribution, as usually required for technical applications. Particle size distributions can only be investigated electron microscopically with the help of some cross-section preparation technology. Multilayer systems made by simultaneous or alternating plasma polymerization and metal evaporation are clearly more appropriate for determining particle size and investigating the interface between metal particles and plasma polymer matrix, because here metal particles are mostly embedded in one plane and films can be investigated using an electron microscope without extra preparation requirements. Because of the two plasma polymer layers on either side, the particles are completely embedded in a homogeneous media which is ideal for determining nanostructural changes. For these investigations, it is thus possible to exclude the influence of substrate material and surface on diffusion processes which change the nanostructure.

Additionally, plasma polymer multilayer systems with embedded metal nanoparticles which have a continuously variable metal content are ideally suited to investigating optical and electrical properties of the films and their relation to the particle size and shape distribution, because many samples can be investigated from such films with identical plasma polymer properties but different metal contents. Almost all investigations presented in this book were carried out on such multilayer systems consisting of a first plasma polymer film, a plasma polymer film with embedded nanoparticles with continuously varying metal content, and a second plasma polymer film. The two plasma

Table 2.1. Deposition and properties of some plasma polymer multilayer systems investigated in the following chapters

Film	Deposition method	Monomer	Nanoparticles	Power density p [W cm^{-2}]	Total thickness d [nm]
I/1	APPME	benzene	silver	0.30	100
I/2	APPME	benzene	silver	0.50	120
I/3	APPME	benzene	gold	0.50	120
I/4	APPME	benzene	indium	0.50	120
I/5	APPME	benzene	silver	0.15	120
I/6	APPME	benzene	silver	0.85	120
I/7	APPME	HMDSN	silver	1.00	260
I/8	APPME	HMDSN	silver	0.15	420
I/9	APPME	styrene	silver	0.85	160
I/10	APPME	HMDSN	copper	1.00	1100
I/11	APPME	benzene	indium	0.50	220
I/12	APPME	benzene	silver	0.03	80
I/13	APPME	benzene	silver	0.85	150
I/14	APPME	HMDSN	silver	0.50	110
I/15	APPME	benzene	silver	0.15	180
I/16	APPME	HMDSN	gold	0.50	150
I/17	APPME	HMDSN	gold	0.85	150
I/18	APPME	HMDSN	gold	0.15	120
I/19	APPME	benzene	silver	0.15	80
I/20	APPME	HMDSN	silver	0.40	60
I/21	APPME	HMDSN	silver	0.40	40
II/1	SPPME	HMDSN	indium	0.50	120
II/2	SPPME	styrene	indium	0.85	150
II/3	SPPME	benzene	silver	0.85	120
II/4	SPPME	benzene	silver	0.85	150
II/5	SPPME	benzene	silver	0.30	120
II/6	SPPME	HMDSN	silver	0.85	260
II/7	SPPME	benzene	silver	0.15	120
II/8	SPPME	benzene	silver	0.60	100
II/9	SPPME	benzene	silver	0.85	80
III/1	PPMS	C_3F_8	gold		200

polymer films usually have roughly the same thickness $d_{p1} \simeq d_{p2}$. The film thickness $d_c(f)$ of the metal-containing film depending on the filling factor is equivalent to the vertical diameter of the largest metal particle and increases with a higher metal content depending on the evaporator distance. This is illustrated in Fig. 2.11 which gives the typical dependence of film thickness on sample position (film I/20 in Table 2.1). Sample position 1 is nearest to the evaporator. The high metal content of the sample nearest to the evaporator is responsible for the substantially higher thickness of the metal-containing part of the multilayer system. The area filling factor F_A was determined by TEM investigations which are described in Chap. 3. For small metal proportions (area filling factor $F_A < 0.65$), the film thickness decreases only slightly and almost linearly. For most of the investigations presented later, films were selected with low area filling factors and approximately constant thickness of the metal-containing layer.

Table 2.1 gives the deposition method, the monomers and metals, the power density p used for plasma polymerisation and the total thickness d of the films investigated in the following chapters. The total thicknesses are given for multilayer systems consisting of three layers, except for film I/10 which consists of five layers and Film III/1 which has a homogeneous vertical particle distribution. An exact determination of $d_c(f)$ is only possible by cross-sectional TEM investigation (Sect. 3.2.2). For metal contents below the percolation threshold f_c, an average thickness d is therefore assumed.

3. Nanostructure

3.1 Characterizing Nanostructure

3.1.1 Nanostructural Analysis

A prerequisite for correlating nanostructure with electrical and optical properties is an extensive determination of the nanostructure itself. The nanostructure of polymer films with embedded metal nanoparticles was defined in Chap. 1 as the lateral and vertical particle size and shape distribution. However, the film surface, inner interfaces and substrate interface must also be considered. This means that, in addition to information about size, shape and distribution of embedded particles, details of the inner interfaces (e.g., formation of particle shells, changes in polymer structure), the film surface, the interface between film and substrate, and the structure of the basic elements (e.g., grain boundaries within the metal particles) are also needed to determine the nanostructure.

For methodological reasons, it is useful for these thin film structural investigations to distinguish between lateral and vertical film structures (see Fig. 3.1). Lateral nanostructure can be determined using transmission electron microscopy (TEM). From TEM bright-field images, the size and shape of embedded particles can be determined and the interfaces between metal particle and plasma polymer investigated. The crystal structure, for example, can be determined from electron diffraction images. This is also possible using high resolution electron microscopy (HRTEM), where grain boundaries and dislocations in the crystal structure of metal particles can also be observed. Furthermore, statements about crystal orientation are possible from dark-field images.

With cross-sectional preparation of the films, vertical nanostructure can also be determined by transmission electron microscopy. The vertical size and shape of embedded metal particles and their spatial distribution in the plasma polymer matrix can be determined using this method. Likewise inner lateral interfaces, for example, the transition between two different plasma polymer morphologies, and film substrate interfaces can also be analyzed.

Scanning electron microscopy (SEM) and atomic force microscopy (AFM) can be used as image-producing analysis methods for examining the film sur-

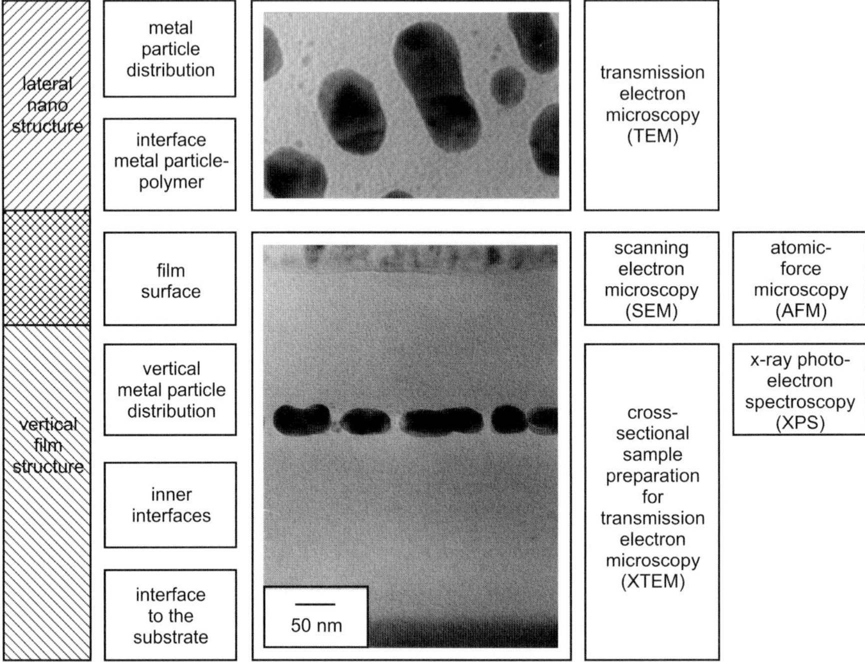

Fig. 3.1. Determination of the lateral and vertical nanostructure of a multilayer system containing metal nanoparticles

face. Both provide direct imaging, and imaging of the film surface at atomic resolution is in principle possible with atomic force microscopy.

X-ray photoelectron spectroscopy (XPS) allows for analysis of the chemical elements in the film, apart from hydrogen. Information about chemical bonds on the surface can be ascertained from the chemical shift in binding energies, so that possible adsorbents can be detected. Furthermore, by film etching, a depth profile can be determined for the different binding energies. Information about the vertical distribution of the metal particle in the plasma polymer matrix can be given from the depth profile, and possible modifications in the chemical structure of the polymer matrix can be detected.

Discussions of nanostructure in this book are limited to results obtained using the analytical methods introduced rather briefly here. Additional nanostructural studies of polymers with embedded nanoparticles are possible using Auger electron spectroscopy (AES), which also determines the depth profile for a metal particle distribution like XPS [328], or using small angle neutron scattering (SANS), which can also determine the sizes of embedded metal particles [342]. These methods, as well as X-ray diffraction or Rutherford back-scattering, will not be considered in the following.

3.1.2 Sample Preparation

For investigation of polymer films with embedded nanoparticles it may be necessary to carry out sample preparations which are specific to the chosen analytical method. For this reason it is important to avoid modifications of the nanostructure of the film during the required sample preparation.

For investigations with transmission electron microscopy, the film must be placed on the copper grids usually used. A direct deposition on the copper grid is not possible without a carrier layer, because no closed plasma polymer film will grow over the grid gaps. However, copper grids can be used as substrates if an amorphous carbon film is deposited as a carrier layer. For direct deposition on the copper grids, a special fixture is necessary, but deposition conditions change little with this fixture, or with the different substrate material (compared with a glass substrate). It has been observed that films made by simultaneous plasma polymerization and metal sputtering on copper grids deposited with carbon show different metal amounts from those on quartz substrates.

On the other hand, it is possible to deposit the film on a special substrate, dissolve it and put the film on the copper grids. This method has the advantage that on these film samples, optical and, in principle, also electronic measurements are possible before placing the films on the copper grids. Silicon wafers or glass/quartz substrates predeposited with potassium bromide (KBr) or sodium chloride (NaCl) are very effective substrate materials for such sample preparation. Since the thin (10–50 nm) KBr film dissolves immediately in distilled water, the film detaches from the substrate, and can be caught on the copper grids. This is an easy sample preparation method for TEM investigations. Investigations with the light microscope and SEM showed that folding and overlapping could occur on these prepared samples, but a sufficiently large area could still be found where the film retained its original condition. No changes in metal particles could be found for this kind of sample preparation when placing the polymer films in distilled water.

In order to receive bright-field images in the transmission electron microscope, transmissible substrates are needed. Because of contrast reduction from the plasma polymer layer, the film thickness of the multilayer systems should be in the range 100–200 nm for a 200 keV transmission electron microscope. The multilayer systems described in Sect. 2.2.2 are ideal for analysing the lateral size and shape of embedded particles, because single particles in the films can be depicted clearly. The determination of vertical structure already mentioned requires a special preparation in order to get a transmissible thin vertical film structure. Cross-sectional sample preparation techniques can be carried out in the conventional way using mechanical thinning and ion beam etching and using the so-called focused ion beam technology (FIB).

For conventional preparations, the polymer film with embedded metal nanoparticles is deposited on a silicon wafer. A thin aluminum layer is deposited by thermal evaporation to protect the surface of the top layer of the plasma polymer film. Two small pieces of the wafer are glued face to face and thinned to a thickness of about 30 μm by mechanical polishing and chemical grinding. After that, samples are ion milled from both sides at low incidence angles (5°) to the surface using two ion guns with 3–8 kV acceleration voltage. It is important to reduce the acceleration voltage at the end of the thinning procedure in order to avoid thermal loading of the samples, and especially of the organic layers. The entire vertical structure of the plasma polymer multilayer system can be investigated with the TEM using these specially prepared, cross-sectional samples.

Focused ion beam technology (FIB) gives a tool for high-precision cross-sectional preparation and surface investigation during preparation as well. A focused ion beam workstation operates as an ion scanning microscope with a liquid metal ion source. The focused gallium ion beam scanned over the specimen generates secondary electrons and ions from the surface and can be used for imaging and physical sputtering with submicrometer resolution. Various gas injection systems allow chemically assisted selective etching of metals or insulators and chemical vapor deposition of platinum or silicon oxide, for example. The FIB technique can thus be used for maskless micro- and submicro-machining of specimens. The FIB technique is especially useful for preparing precisely positioned cross-sections for SEM and TEM.

No special sample preparation is needed For photoelectron spectroscopy (XPS) and atomic force microscopy (AFM). Films deposited on a substrate (e.g., silicon wafer) can be investigated without any pretreatment. A conductive film surface (e.g., platinum layer) is useful if high resolution scanning electron microscopy investigations (HRSEM) will be performed.

3.1.3 Determination of Particle Geometry

In order to determine the nanostructure of plasma polymer films with embedded metal nanoparticles, it is necessary to quantify electron microscopy results concerning the size and shape of embedded particles. Commercial image processing systems can be used to evaluate TEM micrographs. These can be directly stored as computer graphic images from the transmission electron microscope, or the micrograph negative can be scanned by a CCD camera, and the results displayed on a monitor and digitized by a frame-grabber. The digitized TEM micrographs are then evaluated by means of an image analysis software (see Fig. 3.2).

Two algorithms can be applied to characterize digitized particles. In the first, the area center of gravity of the particle is determined and 36 angle sections (ferets) are drawn through it. Single particle diameters are then calculated. The other algorithm proceeds by lining up a sufficient number of tangents against the circumference of the particle. Both ways lead to an almost exact determination of the particle area A and circumference P, as well as the largest particle diameter $D_{\max}$ and the smallest particle diameter $D_{\min}$. The particle size D is defined as the diameter of an equivalent circular

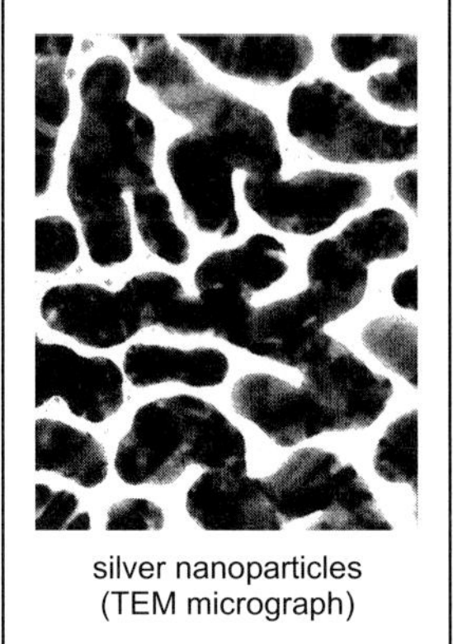 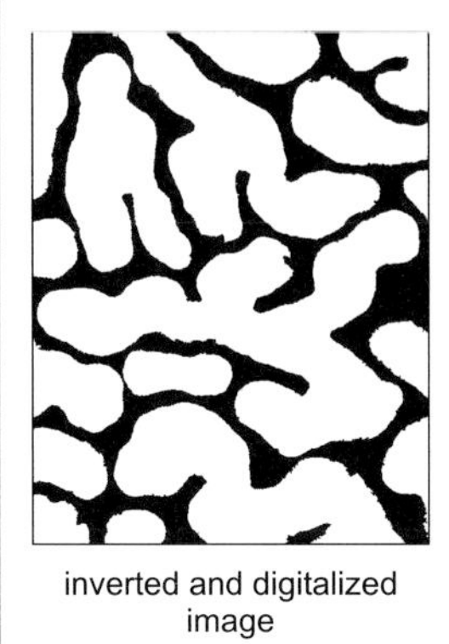 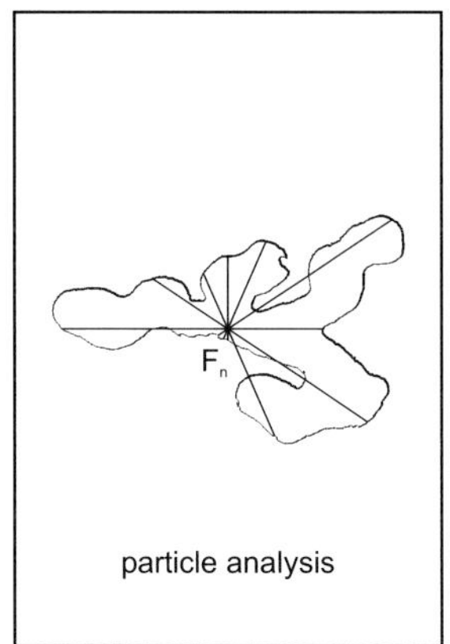

silver nanoparticles (TEM micrograph) | inverted and digitalized image | particle analysis

Fig. 3.2. Image processing and image analysis of TEM micrographs

area. The irregular silver particle depicted in Fig. 3.2 has area $A = 7\,560$ nm^2, particle circumference $P = 652$ nm, largest particle diameter $D_{\mathrm{max}} = 128$ nm, and smallest particle diameter $D_{\mathrm{min}} = 113$ nm. The particle size is $D = 98$ nm.

From these values the particle shape factor S and the aspect ratio Q can be calculated for every single particle with $S = 4\pi A/U^2$ and $Q = D_{\mathrm{min}}/D_{\mathrm{max}}$. The values for the particle shape factor S and the aspect ratio Q lie in the range $0 \le S \le 1$ and $0 \le Q \le 1$, where $S = 1$ and $Q = 1$ correspond to a circular object. For most small, almost ideally circular particles, typical values for S are around 0.9 by calculation from the digitally determined quantities A and U. In particular, for very irregular particles, the shape factor S gives more information for the characterization of a particle than the aspect ratio Q. For example, the particle presented in Fig. 3.2 has shape factor $S = 0.22$ and aspect ratio $Q = 0.88$. Note that this aspect ratio of $Q = 0.88$ would suggest an almost circular object.

Furthermore, an area filling factor F_{A} can be determined from the black-to-white ratio after the transition from an analog image to a digitized one, where particles correspond to white areas and plasma polymer to black ones (see Fig. 3.2). However, it must be pointed out that the area filling factor F_{A} is not equivalent to the volume filling factor f discussed in Chap. 1, and only with information about the vertical particle diameters are any conclusions possible as to the volume content of metal in the composite film.

After image analysis of the TEM micrograph, it is possible to carry out statistical calculations from the number of metal particles (30–500). Average values can be calculated for the particle area $\bar{A}$, the particle circumference $\bar{U}$, the largest particle diameter $\bar{D}_{\mathrm{max}}$, the smallest particle diameter $\bar{D}_{\mathrm{min}}$, the mean particle size $\bar{D}$, the shape factor $\bar{S}$ and the aspect ratio $\bar{Q}$. Histograms can then be drawn up for the particle size and particle shape distributions. A combined histogram showing particle sizes and shape factors in a three-

dimensional diagram is very useful in order to understand the particle size
and shape distributions more clearly. The area of each particle is plotted (z
axis) on a grid determined by the particle size (x axis) and the shape factor
(y axis). These three-dimensional particle size and shape histograms are used
for further discussion of the metal particle size and shape distribution. The
geometric quantities for single particles and particle systems which are useful
when discussing TEM micrographs are:

- single particle:
  - particle size D,
  - particle shape factor S,
- particle systems:
  - mean particle size with standard deviation $\bar{D} \pm \sigma_D$,
  - mean particle shape factor with standard deviation $\bar{S} \pm \sigma_S$,
  - area filling factor F_A,
  - three-dimensional particle size and shape histograms.

It should be noted that the results of the image analysis are influenced by
several subjective factors. Image enhancement steps such as the gray value
filter, multiple addition of the same image and others, required to improve
contrast between particles and areas without particles and to transform the
analog micrograph into a digital image, influence the results of subsequent
image analysis. Different functions are used to reduce the roughness of particle edges, and can influence results for the shape factor. For example, two
neighboring particles can occur in the digitized image as one object and
must be separated manually. Furthermore, particles which are cut by the
edge of the micrograph and single gray pixels which are produced by digiti-
zation must be removed. These subjective factors can be minimized by using
a standard procedure (macro programming) for all manual and arithmetic
operations. This procedure guarantees comparability of all image analysis results. It should also be mentioned that whole TEM m icrographs were used
for the investigations, but due to lack of space only sections are shown in the
figures.

3.2 Particle Size and Shape Distribution of Embedded Metal Nanoparticles

3.2.1 Particle Size and Shape Distributions in the Lateral Direction

Various Embedded Metal Nanoparticles. In order to determine the
exact size and shape of embedded metal particles, the objects must be
clearly separated in electron microscope micrographs. Since metal-particle-
containing polymer films with a vertical homogeneous distribution have particles one on top of the other, single objects cannot be separated from one

another, so that particle size and particle shape analysis is almost impossible. Consequently, multilayer systems have been used for the investigations, as explained in Sect. 2.2.2. These are made by simultaneous or alternating plasma polymerization and metal evaporation.

As already mentioned in Sect. 2.2, an island growth process (Volmer–Weber growth) is expected for the evaporated metals under the chosen conditions for depositing the plasma polymer film with embedded nanoparticles (i.e., substrate at room temperature, absolute pressure in the reactor $\approx$ 1 Pa). If sufficient metal atoms are available, an almost continuous metal film can grow. Samples of plasma polymer multilayers with continuously varying metal particle size and shape, taken in sections with different metal content, exhibit all three structure ranges possible for insulator–metal composite films (Chap. 1).

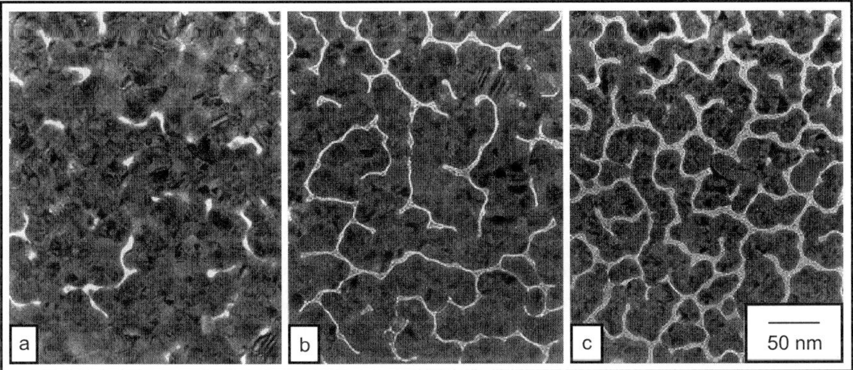

Fig. 3.3. (a–c) TEM bright-field images of three samples of a metal-containing plasma polymer multilayers with continuously varying silver content

Figure 3.3 shows three TEM[1] bright-field images of a multilayer system with a plasma polymer film (film I/1) with three different metal contents. Figure 3.3a shows a nearly closed metal layer placed between two plasma polymer layers (metallic structure range). This metal layer consists of many randomly oriented silver crystallites, and these generate continuous diffraction rings with brighter spots for single large particles in the diffraction pattern. Figure 3.3b shows a sample with a nanostructure at the percolation transition f_c (percolation structure). The silver layer is not completely closed, but no particles are available that are isolated from each other. Figure 3.3c shows single particles which are separated from each other (insulating structure range). Diffraction rings will be found for these larger particles in the diffraction

---

[1] TEM investigations were carried out with a JEOL 100 CX at 100 keV or with a Philips CM 20 at 200 eV

pattern. TEM dark-field images have shown that larger ($D \geq 10$ nm) embedded silver particles are polycrystalline, and single crystallites are randomly orientated (Fig. 3.18).

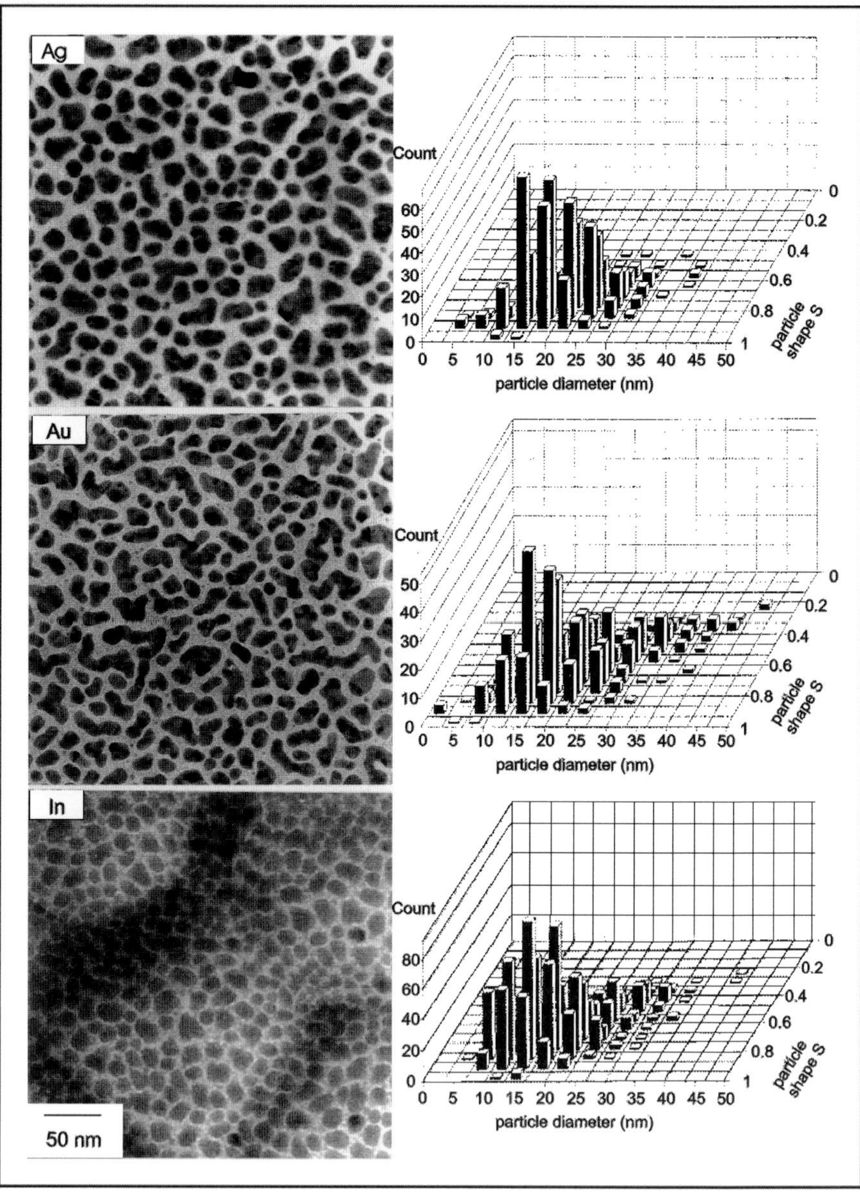

Fig. 3.4. TEM bright-field images and three-dimensional size and shape histograms for plasma polymer multilayers with embedded silver, gold and indium particles

The embedding of polycrystalline particles also takes place in plasma polymers with other metal nanoparticles. Figure 3.4 shows TEM micrographs for plasma polymer multilayers with (a) embedded silver nanoparticles (film I/2), (b) embedded gold nanoparticles (film I/3) and (c) embedded indium nanoparticles (film I/4), as well as the three-dimensional particle size and shape histograms for these films constructed from image analyses. Samples with comparable mean particle sizes were chosen from these three films. The mean particle sizes and shape factors were determined as $\bar{D} = 20.6 \pm 5.3$ nm and $\bar{S} = 0.81 \pm 0.08$ for embedded silver particles, $\bar{D} = 20.4 \pm 6.9$ nm and $\bar{S} = 0.72 \pm 0.14$ for embedded gold particles, and $\bar{D} = 18.0 \pm 5.8$ nm and $\bar{S} = 0.75 \pm 0.13$ for embedded indium particles.

Bright-field TEM images of the silver and gold particles look very similar. While no particles exist in the silver sample with particle sizes $D > 40$ nm and shape factors $S < 0.4$, there are still very few particles with $D > 40$ nm and $S < 0.4$ in the gold sample, which leads to a higher standard deviation of the particle size and a lower mean shape factor. Although all three samples have similar particle sizes, the indium particles are formed differently. No long, stretched particles were observed, such as those seen in the gold sample. The similar value for the indium and gold mean shape factors is caused by the fragmentary shape of the indium particles. Dark regions in the TEM micrographs result from the surrounding plasma polymer matrix. The influence of the polymer matrix on particle size and shape distribution will be discussed in the next chapter.

The embedding of polycrystalline particles has been detected for silver, gold and indium, and also for bismuth, tin and copper [336]. The formation of polycrystalline particles has not been observed for embedded holmium [309]. The amorphous electron diffraction pattern shows that a chemical reaction takes place between holmium and the monomer or the residual gas in the reactor, leading to a reduction in monomer pressure when holmium is evaporated. Both amorphous and polycrystalline particles have been observed for embedded germanium [309]. The embedding of selenium is also amorphous, but very large selenium crystallites ≥ 500 nm are also formed.

Influence of the Plasma Polymer on Particle Size and Shape Distributions. Since plasma polymer films with embedded metal particles are usually made in the form of multilayer systems, so that metal evaporation starts after or during plasma polymerization, the growth process of the metal particles is determined by surface properties of the plasma polymer before deposition or during its formation. Moreover, a special plasma polymer morphology (cauliflower morphology) which occurs with increasing film thickness (see Sect. 2.1.3) influences the size and shape distribution of embedded metal particles.

Figure 3.5 shows TEM micrographs for two multilayer systems without this plasma polymer morphology (film I/5, Figs. 3.5a, c and e) and with a strongly formed plasma polymer morphology (film I/6, Figs. 3.5b, d, and f).

Dark regions in Fig. 3.5b result from the high concentration of embedded silver particles, as shown in Fig. 3.5d.

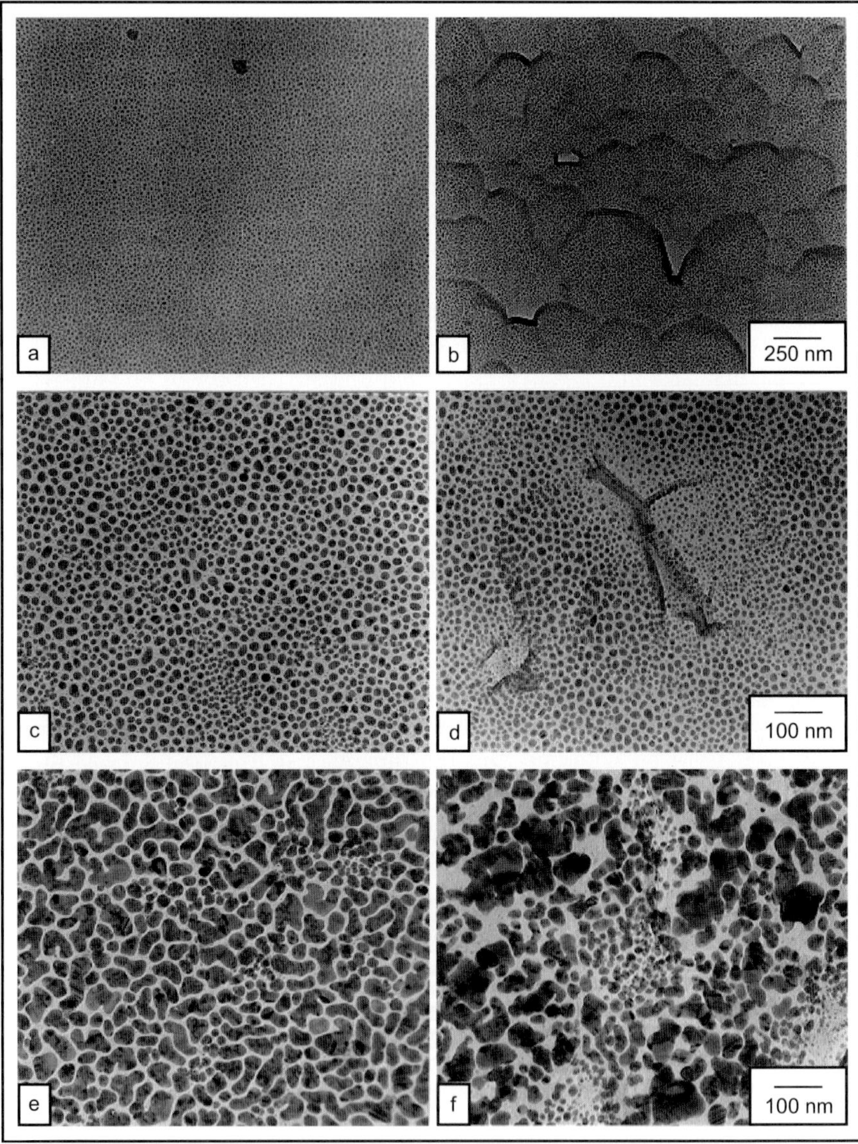

Fig. 3.5. (a–f) Influence of plasma polymer morphology on the particle distribution in plasma polymer thin films with embedded silver nanoparticles

The influence on particle size, already visible in this picture, is depicted by means of the TEM micrographs for a sample of the film with a higher metal content (Fig. 3.5f). Particles are larger in the undisturbed regions and smaller and more numerous in the disturbed regions.

This specific particle distribution is clearly shown in Fig. 3.6 which gives the particle size and shape histograms for the images in Figs. 3.5e and f. The mean particle size and the mean shape factor in those cases are $\bar{D} = 39.4 \pm 17.4$ nm, $\bar{S} = 0.61 \pm 0.20$ for Fig. 3.5e and $\bar{D} = 25.1 \pm 23.7$ nm, $\bar{S} = 0.68 \pm 0.22$ for Fig. 3.5f. While no particles with $D > 100$ nm occur in the polymer matrix without surface morphology (upper image), a great many silver particles with $D > 100$ nm can be found in the polymer matrix with surface morphology (lower image, cauliflower morphology). The lower histogram of Fig. 3.6 also shows the large number of very small silver particles, with $D < 12$ nm, that causes a significantly lower mean particle size compared to the particles in Fig. 3.5e, and also a larger standard deviation.

The differences between the particle size distributions can be explained by a simple model for the growth behavior of metal particles that considers the reactor setup for fabricating metal-containing plasma polymer films (see Fig. 3.6). The cauliflower morphology causes hills and valleys in the plasma polymer surface. Since the substrate is located on the lower electrode (Fig. 2.10), the deposition angle of the metal atoms is not vertical. For this reason, the valley walls facing the evaporator witness the growth of more particles than the shadow sides of the valleys. Surface diffusion of impacting metal atoms can take place undisturbed at the top of the hills, so that fewer but larger particles form.

The size differences of particles on hills and in valleys confirm the assumption that this special particle distribution is caused by the polymer film morphology and does not result from the film preparation for electron microscopy (see Sect. 3.1.2). These specific regions of the particle distributions make it easier to recognize sample regions in the electron microscope.

The typical metal particle distribution caused by the cauliflower morphology has also been observed for the embedding of other metals (gold, copper), and occurs especially for multilayer systems in which the first plasma polymer film has a relatively large thickness d_{p1} and was deposited at relatively high power densities. During metal evaporation, the surface of the lower plasma polymer layer is decorated in the same way as an inner interface and so the plasma polymer morphology is visible in the TEM micrographs.

While no significant differences were found for gold and silver embeddings with comparable deposition conditions, a strong dependence of particle shape on the plasma polymer matrix was observed when indium particles were embedded. Figure 3.7 shows this dependence by means of the TEM micrographs together with three-dimensional particle size and shape histograms for samples of three plasma polymer thin films with embedded indium nanoparticles

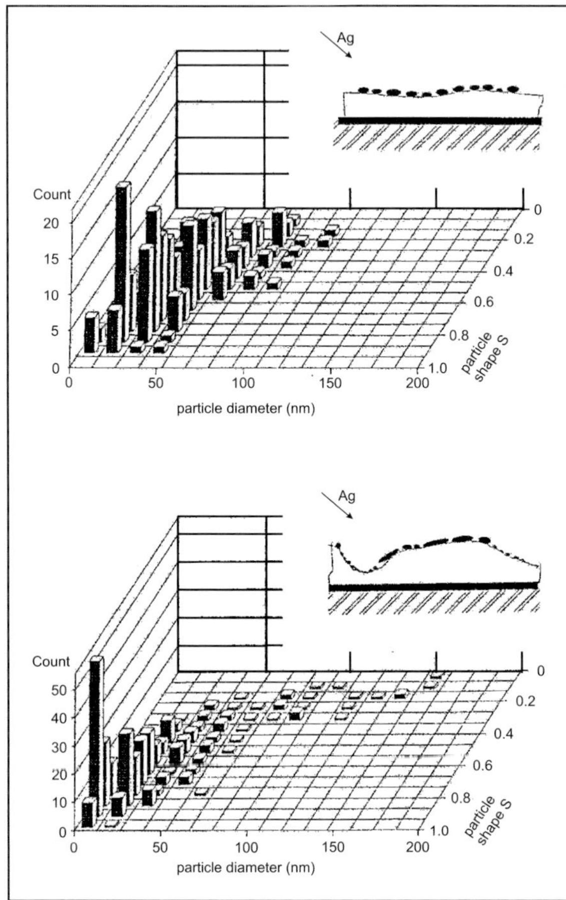

Fig. 3.6. Three-dimensional particle size and shape distribution histograms and film growing models for the TEM micrographs in Figs. 3.5e and f

made from different matrix materials: (a) film I/2, (b) film II/1, and (c) film II/2.

While indium particles are nearly spherical for the samples of Fig. 3.7b and c ($\bar{S} = 0.88 \pm 0.05$ and $\bar{S} = 0.88 \pm 0.04$), a mean shape factor of $\bar{S} = 0.66 \pm 0.12$ was determined for these particles in the sample of Fig. 3.7a. They have concave surfaces, and the gap between two neighboring particles is about 20 nm. Notably for samples (b) and (c), the gaps between larger indium particles are filled with very small indium particles ($D \leq 20$ nm). This leads to a bimodal particle size distribution and to a decrease in the mean particle size: $\bar{D} = 29.4 \pm 22.4$ nm for Fig. 3.7b and $\bar{D} = 29.7 \pm 22.9$ nm for Fig. 3.7c. If only particles with $D > 20$ nm are considered, the mean particle sizes amount to $\bar{D} = 50.9 \pm 12.5$ nm and $\bar{D} = 52.3 \pm 15.1$ nm, respectively, and range close to the values for particles in Fig. 3.7a with $\bar{D} = 65.7 \pm 18.3$ nm.

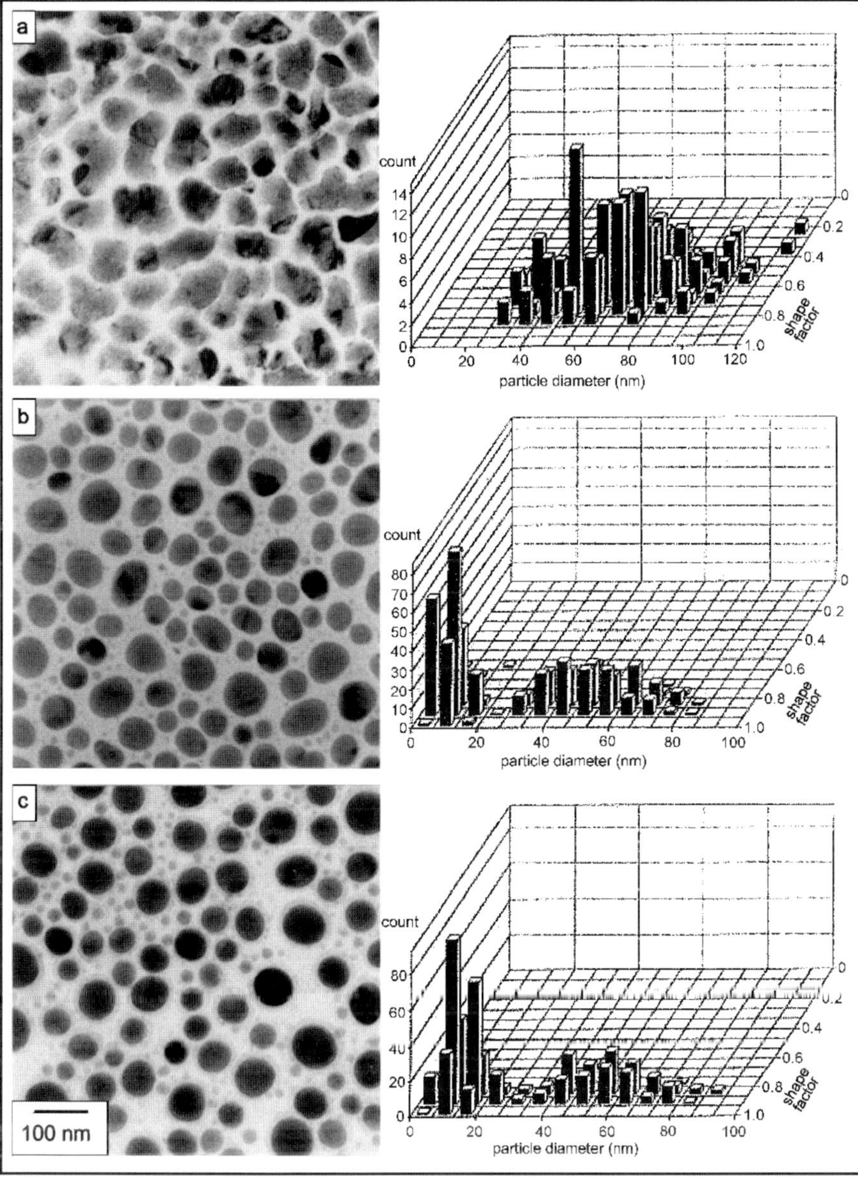

Fig. 3.7. TEM micrographs of three plasma polymer multilayer systems with embedded indium particles made from (**a**) benzene, (**b**) HMDSN and (**c**) styrene

The particle shape presented in Fig. 3.7 forms independently of the cauliflower morphology at higher film thicknesses, and has been found only for plasma polymer films with embedded indium nanoparticles made from benzene with low power density ($p \leq 0.5$ W cm^{-1}) and by alternating plasma polymerization and metal evaporation. For a particle distribution with two clear maxima, the smaller particles are often concentrated around the larger particles. This indium particle distribution has been observed for embedding in HMDSN plasma polymer films, but especially in films made from styrene with higher power densities ($p \approx 0.85$ W cm^{-1}).

The very clear differences in the indium particle shape depending on the monomer species were not observed for other embedded metals. The formation of indium oxide during deposition may explain this effect, but diffraction analysis of all samples in Fig. 3.7 was unable to detect any indium oxide, with a detection limit for In$_2$O$_3$ of about 3% in electron diffraction analyses.

Spherical indium particles ($D \approx 50$–100 nm) have been investigated for indium deposition films on polymer substrates, where fewer small particles with $D \approx 5$–10 nm were found between the large particles [343]. But these small particles are not located in the form of an encapsulation around the larger particles. SEM investigations of indium deposition films on SiO$_2$ [344] or of indium epoxy composite films [170] showed particle size and shape distributions comparable to those for silver and gold particles in Fig. 3.4.

Figures 3.5 and 3.7 both show that the surface morphology of the first plasma polymer film, which still exists before metal particles are embedded, has a decisive influence on particle size and shape distributions of embedded metal particles. The metal particles decorate the surface of the already existing plasma polymer layer, as was also observed for gold particles on polystyrene [345] or silver particles on Langmuir–Blodgett films [346], and consequently provide information about the film surface. In order to distinguish single particles for further investigations of particle size and shape distributions, films without a cauliflower morphology were used.

Particle Size and Shape Distributions for Different Mean Particle Sizes. In this section, the particle size and shape distributions for plasma polymer films with embedded nanoparticles having a continuously varying metal content are discussed. This sets the stage for describing optical properties in Chap. 6, because optical measurements were mainly carried out on these films.

Figure 3.8 shows the TEM micrographs for six samples of a multilayer system (film I/5). This system has a plasma polymer film with embedded silver nanoparticles and a continuously varying metal content. The silver content decreases continuously from Fig. 3.8a to f. The mean particle size $\bar{D}$ decreases with decreasing metal content, and the shape factor $\bar{S}$ decreases with increasing metal content. The standard deviations of the mean particle size σ_D and the mean shape factor σ_S also decrease in correspondence with the mean particle size and mean particle shape. The increase in shape factor

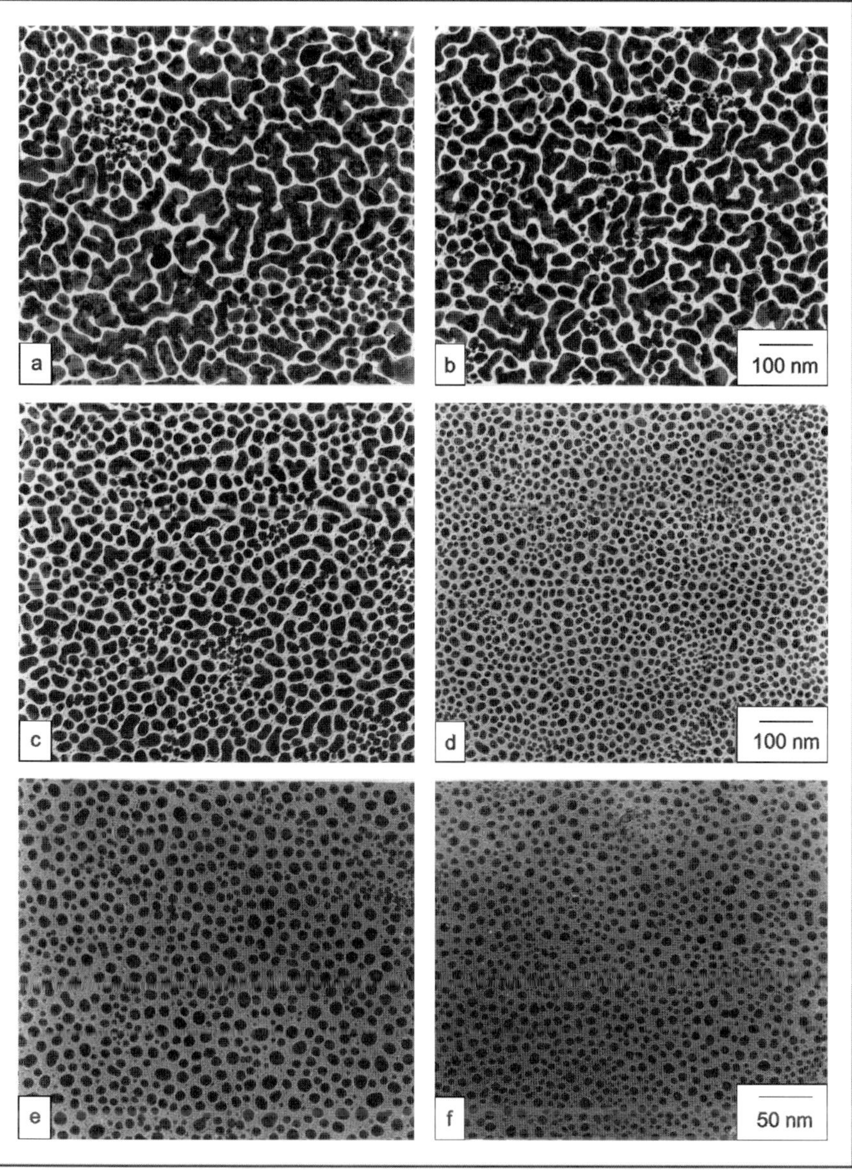

Fig. 3.8. (a–f) TEM micrographs for six samples of plasma polymer films with embedded silver nanoparticles of continuously varying particle size

with decreasing particle size means that smaller metal particles are closer to
a spherical shape.

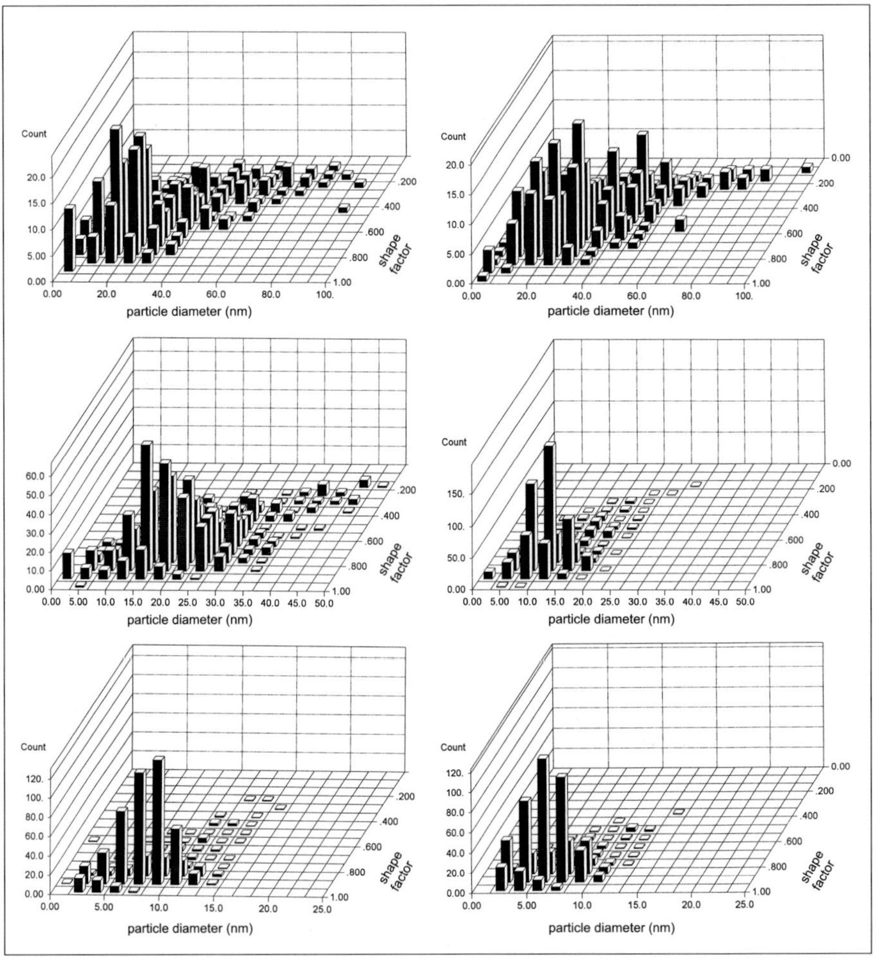

	$\bar{D}$ [nm]	$\bar{S}$		$\bar{D}$ [nm]	$\bar{S}$
(a)	30 ± 19	0.56 ± 0.20	(b)	32 ± 16	0.58 ± 0.19
(c)	20 ± 8	0.70 ± 0.16	(d)	12 ± 3	0.79 ± 0.12
(e)	8.0 ± 2.6	0.86 ± 0.10	(f)	6.2 ± 2.4	0.87 ± 0.09

Fig. 3.9. Particle size and shape histograms for the TEM micrographs in Fig. 3.8,
together with the mean size and mean shape of embedded silver particles

The three-dimensional particle size and shape distribution histograms for the micrographs in Fig. 3.8 are given in Fig. 3.9. It is obvious for samples (a) and (b) that a large number of particles still have particle sizes $60 \leq D \leq 100$ nm and small shape factors $S < 0.4$. There is also a very wide particle size distribution. No particles are found with both a particle size $60 \geq D \geq 100$ nm and a shape factor $S > 0.5$. Only a few particles have shape factors $S < 0.5$ for samples (e) and (f), which have smaller mean particle sizes. The majority of particles are nearly spherical and have shape factors $S \geq 0.8$. Deviations in the mean particle size are significantly lower. There are very few particles whose size is greater than twice the mean particle size.

The particle size and shape distributions with decreasing metal content shown in the TEM micrographs of Fig. 3.8 and the three-dimensional particle size and shape distributions shown in Fig. 3.9 are representative of most plasma polymer multilayer systems with embedded silver nanoparticles. Similar particle size and shape distributions have also been observed for embedded gold particles. The progression of the TEM images from Fig. 3.8a to f is similar to what is observed in TEM micrographs for different stages of growth of evaporated discontinuous metal films (e.g., Ag or Au films [347–350]).

A significantly different progression of particle size and shape distribution for decreasing metal content is observed in a plasma polymer multilayer system (film I/4) with embedded indium nanoparticles that has a continuously varying metal content. As can be seen from the TEM micrographs in Fig. 3.10, very large fragmented particles are likely to occur with a higher metal content, and these do not possess the same narrow shapes as silver or gold particles with higher metal content (Fig. 3.3).

The occurrence of this specific particle shape has already been discussed when describing Fig. 3.7. It is interesting to note that, among the information about the partly concave surface of many indium particles, the indium embedded particles of the samples in Fig. 3.10a and b are much bigger than the particles shown in Fig. 3.7. Embedded silver or gold particles of comparable sizes ($D > 200$ nm) have only been observed in plasma polymer films after thermally induced coalescence (see Fig. 4.5 in Sect. 4.2 and Fig. 4.25 in Sect. 4.4.3).

Figure 3.11 shows size and shape distribution histograms for samples (c), (d), (e) and (f) of Fig. 3.10. Because of the small number of particles, the histograms for samples (a) and (b) were not considered. As for embedded silver and gold particles, the mean particle size $\bar{D}$ and mean shape factor $\bar{S}$ for plasma polymer films with embedded indium nanoparticles also increase with decreasing metal content. However, compared with the embedded silver particles, indium particles show a more uniform particle size, as can be seen from the lower standard deviation σ_D from the mean particle size (Fig. 3.11).

Figures 3.8 and 3.10 show that the mean particle size $\bar{D}$ decreases and the mean shape factor $\bar{S}$ increases with decreasing metal content. The size

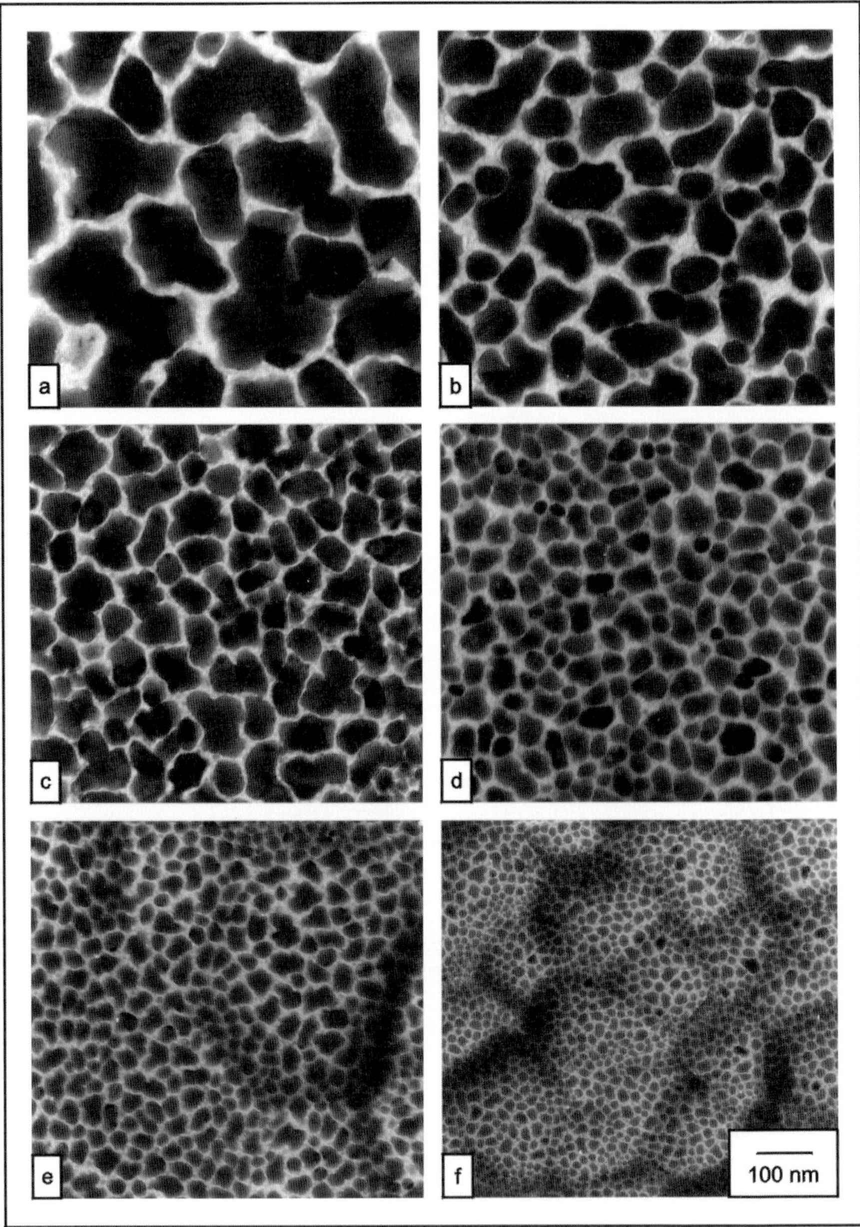

Fig. 3.10. (a–f) TEM micrographs for six samples of a multilayer system of a plasma polymer film with embedded indium nanoparticles and continuously varying particle size

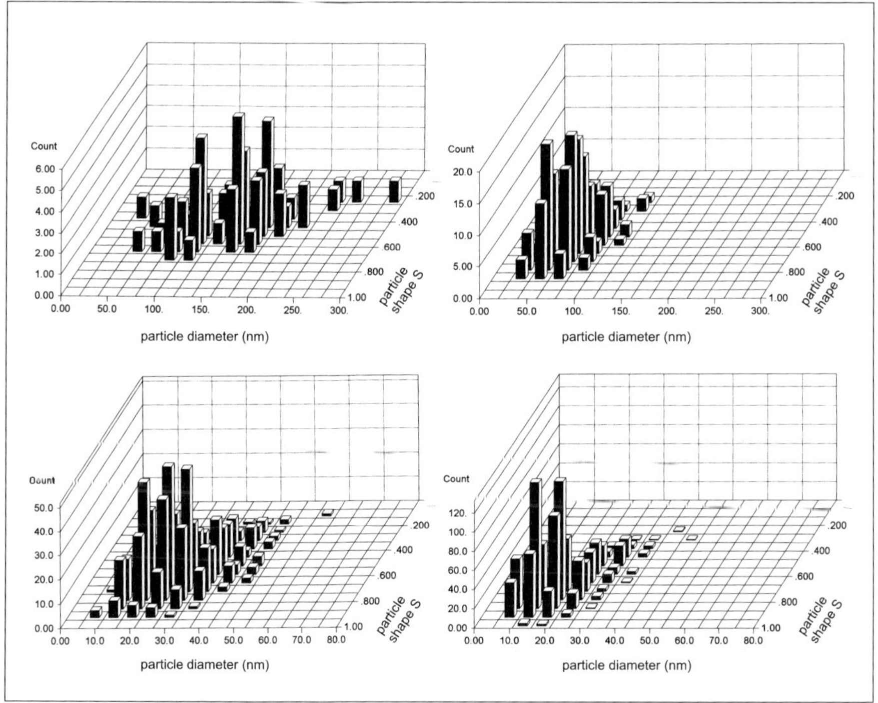

	$\bar{D}$ [nm]	$\bar{S}$		$\bar{D}$ [nm]	$\bar{S}$
(a)	258 ± 143	0.35 ± 0.17	(b)	204 ± 58	0.48 ± 0.11
(c)	143 ± 48	0.51 ± 0.11	(d)	66 ± 17	0.65 ± 0.13
(e)	26 ± 8	0.65 ± 0.15	(f)	18 ± 6	0.75 ± 0.13

Fig. 3.11. Particle size and shape histograms from the TEM micrographs of Fig. 3.10c to f, together with the mean size and mean shape of the indium particles

and shape distributions of Fig. 3.8 are the basis for discussing particle size- and shape-dependent optical properties in Chap. 6.

Logarithmic Normal Distribution. From statistical considerations regarding volume change during growth onto a substrate, a logarithmic normal distribution has been found for the size of spherical particles in [351]. The particle size distribution $f_{\mathrm{LN}}(D)$ is calculated from

$$f_{\mathrm{LN}}(D) = \frac{1}{\sqrt{2\pi}\ln\sigma_{\mathrm{LN}}} \exp\left[-\frac{(\ln D - \ln\bar{D}_{\mathrm{LN}})^2}{2(\ln\sigma_{\mathrm{LN}})^2}\right] , \qquad (3.1)$$

where the quantity $\ln\bar{D}_{\mathrm{LN}}$ is given by

$$\ln\bar{D}_{\mathrm{LN}} = \frac{\sum_i n_i \ln D_i}{\sum_i n_i} \qquad (3.2)$$

and $\ln\sigma_{\mathrm{LN}}$ is calculated from

$$\ln\sigma_{\mathrm{LN}} = \sqrt{\frac{\sum_i n_i(\ln D_i - \ln\bar{D}_{\mathrm{LN}})^2}{\sum_i n_i}}\,. \tag{3.3}$$

The logarithmic normal distribution is used to describe the size distribution of many different metal particle systems that are independent of the kind of metal or the fabrication process, for example Al, Mg, Zn and Sn particles evaporated in an inert gas atmosphere [351], Au particles evaporated in high vacuum [347, 351, 352] or indium particles evaporated in high vacuum [353] and sputtered Ag particles [354] or Cu, Pd and W particles [355]. Agreement is generally very good for mean particle diameters smaller than $\bar{D} = 10$ nm. If there are many small particles between the larger particles, two maxima occur in the size distribution (e.g. Au particles [352]).

Figure 3.12 shows the particle size histograms and logarithmic normal distribution determined by (3.1) for the embedded silver particles of Fig. 3.8. The particle size histograms are based on the three-dimensional particle size and shape histograms shown in Fig. 3.9. The particle size distribution of embedded silver particles for the sample shown in Fig. 3.12d agrees particularly well with the logarithmic normal distribution.

Values for $\bar{D}_{\mathrm{LN}}$ and σ_{LN} determined from the logarithmic normal distribution are shown in Table 3.1.

Table 3.1. Values for $\bar{D}_{\mathrm{LN}}$ and σ_{LN} determined from the logarithmic normal distribution for the TEM micrographs in Fig. 3.8

	$\bar{D}_{\mathrm{LN}}$	σ_{LN}		$\bar{D}_{\mathrm{LN}}$	σ_{LN}
(a)	25.5	1.7	(b)	28.2	1.7
(c)	19.3	1.5	(d)	11.3	1.3
(e)	7.4	1.5	(f)	5.8	1.5

The difference between the values for $\bar{D}$ in Fig. 3.9 and those for $\bar{D}_{\mathrm{LN}}$ in Table 3.1 results from the logarithmic determination of the mean values using (3.2). The values determined for σ_{LN} for the samples (c) to (f) range over $1.3 \leq \sigma_{\mathrm{LN}} \leq 1.6$ for a discontinuous evaporated film [351].

Higher values for σ_{LN} have been determined for larger particles in samples (a) and (b), whereas the logarithmic normal distribution is generally inapplicable for systems with larger metal particles because such particles deviate too much from the spherical shape [351]. The non-ideal polymer surface and possible cauliflower morphology for higher film thicknesses mean that deviations from the logarithmic normal distribution are also expected, since modifications in the growth of the metal particle due to surface roughness of the substrate, or due to competing growth by simultaneous plasma polymerization and metal evaporation are not considered.

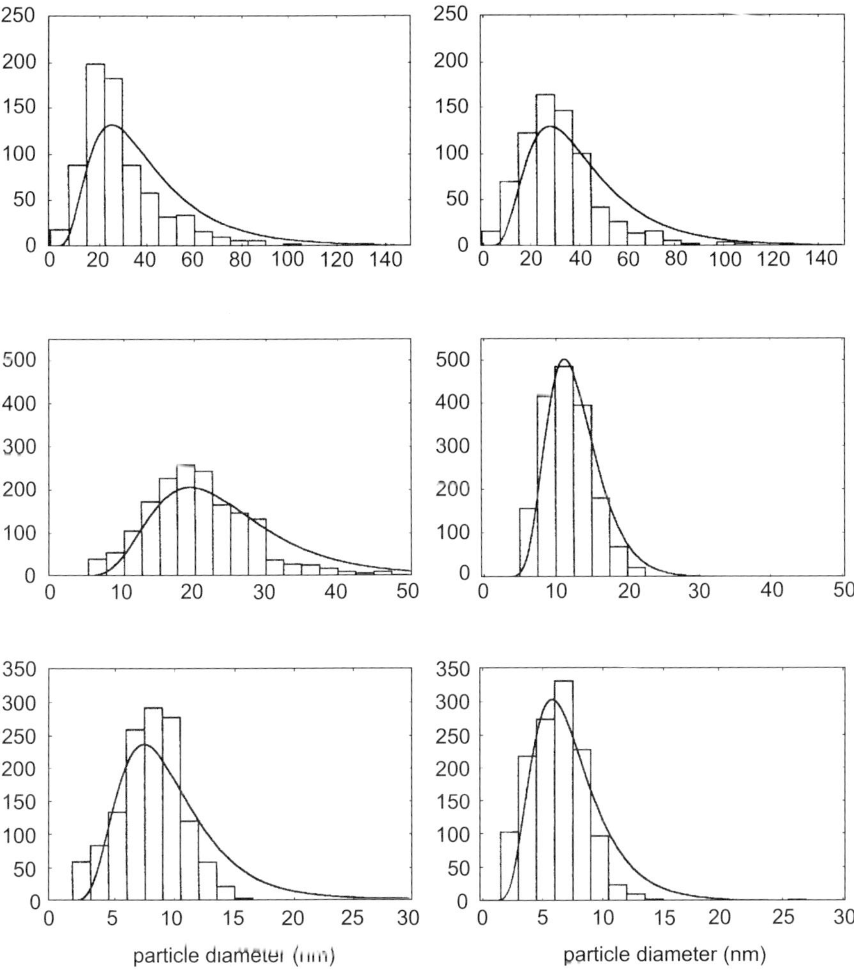

Fig. 3.12. Particle size distribution histogram and logarithmic normal distribution $f_{\mathrm{LN}}(D)$ for embedded silver particles in Fig. 3.8

Different particle size distribution functions can be obtained from thermo-dynamic derivations of thermally induced atomic diffusion in mixed systems (e.g., Ostwald ripening, LSW theory). The particle size distribution is described in Sect. 4.2.1 by means of these particle distribution functions for the plasma polymer films with embedded metal nanoparticles for which modifications in the nanostructure have been brought about by thermal changes.

3.2.2 Particle Size and Shape Distribution
in the Vertical Direction

**Particle Size and Shape Distribution for Cross-Sectional Samples
(XTEM).** TEM microscopy on cross-sectionally prepared samples (see Sect.
3.1.2) gives information about the vertical distribution, and the size and
shape of metal nanoparticles. Figure 3.13 shows such XTEM micrographs
for a multilayer system with embedded silver nanoparticles (film I/7). The
film was made by alternating plasma polymerization and metal evaporation
and shows the expected silver nanoparticles embedded in a plane. The total
film thickness of the multilayer system $d = 260$ nm and the film thickness of
the individual layers $d_{p1} = 120$ nm, $d_c = 40$ nm and $d_{p2} = 100$ nm are all
determined. The plasma polymer shows no morphology in this case.

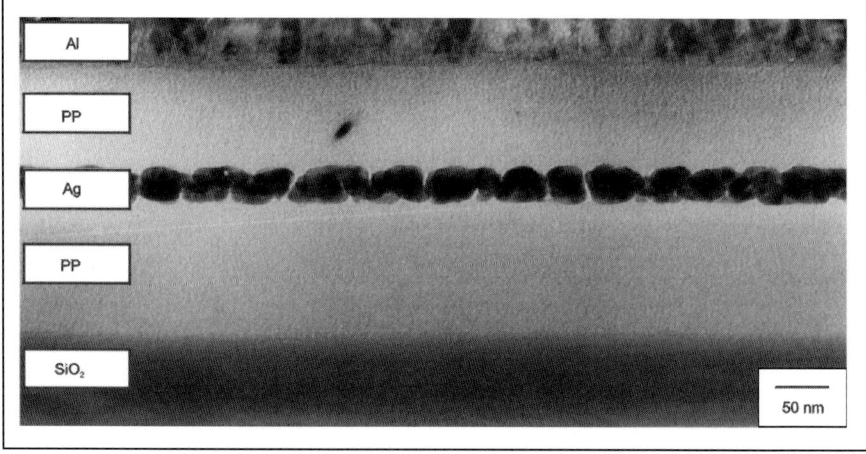

Fig. 3.13. XTEM micrograph of a plasma polymer multilayer system with silver
nanoparticles embedded in a single plane

A different particle distribution results for a cross-sectional preparation
of a plasma polymer multilayer system with embedded silver nanoparticles
made with lower power density (Fig. 3.14, film I/8).

It seems that in some sample regions up to six silver particles are embedded on top of one another, although the plasma polymer does not grow
during metal evaporation. Consequently, the actual composite layer between
the two plasma polymer layers seems to have a partly three-dimensional particle distribution. These specific regions, where embedded silver particles seem
to occur one on top of the other, may be influenced by substrate impurities
(arrow in Fig. 3.14) which cause a small hillock on the surface of the first
plasma polymer layer. The particles are deposited around and on top of this
hillock and at different distances from the substrate. If particles are deposited

around these hillocks, it seems in the vertical view as though the particles are deposited on top of each other, which is not the case for the described sample. Another explanation for this effect could be that the cauliflower morphology of the lower plasma polymer layer already exists before metal evaporation, and that the formation of metal particles is influenced by the formation of hillocks and trenches (Fig. 3.6).

The film thickness of the single layers, especially for the metal-particle-containing layer, can only be estimated owing to the non-uniform lateral embedding of the silver particles. The film thickness is $d_{p1} = 180$ nm for the lower plasma polymer layer, $d_c = 30\text{--}60$ nm for the metal–containing layer and $d_{p2} = 180\text{--}210$ nm for the upper plasma polymer layer. This implies a total film thickness of $d = 420$ nm for the multilayer system. Localized enlargements of the metal–containing layer thickness d_c caused by an inhomogeneous substrate surface or the cauliflower morphology proceed through the second plasma polymer layer and lead to an increase in film surface roughness.

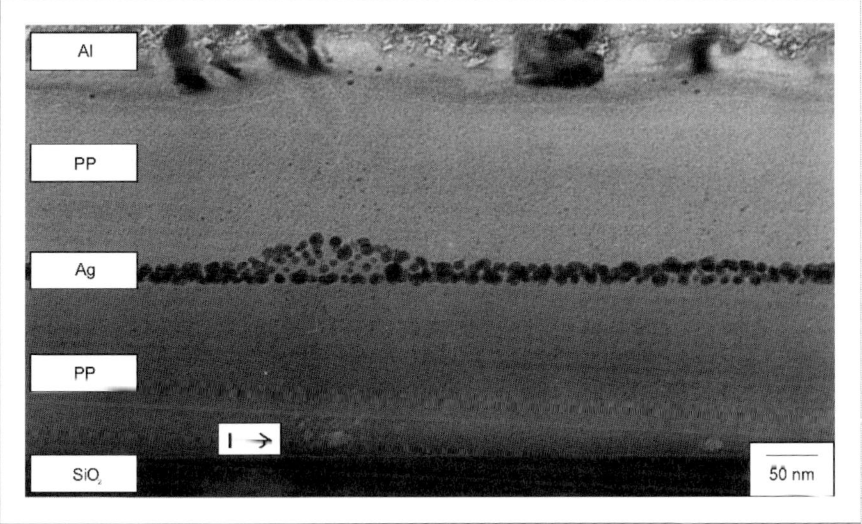

Fig. 3.14. XTEM micrograph of a plasma polymer multilayer system with embedded silver nanoparticles

Figures 3.13 and 3.14 show XTEM micrographs which provide additional information about spatial particle distributions that could not be determined by lateral transmission electron microscopy. It is possible to carry out vertical imaging of single particles by cross-sectional sample preparation, and image analysis can thus be carried out as for lateral TEM micrographs (see Fig. 3.17 in Sect. 3.2.3).

Determination of Vertical Particle Distribution by X-ray Photoelectron Spectroscopy (XPS). Depth profiles can be determined for plasma polymer multilayer films with embedded metal nanoparticles using X-ray photoelectron spectroscopy. These studies are carried out by stepwise argon etching of the layer with a vertical resolution of a few nanometers. This yields no information about the size of embedded particles, but investigation of the vertical metal particle distribution is possible with much less preparation than cross-sectional samples.

A wide range of argon etching rates must be considered for metals and plasma polymers when analysing depth profiles. The experimentally determined etching rates of the plasma polymers (see Sect. 2.1.3) are usually lower at 450–150 Å min^{-1} than the metal etching rates (1 100–1 300 Å min^{-1} for Ag, 1 000–1 400 Å min^{-1} for Au), for Ar etching with 500 V acceleration voltage and vertical incidence angle [127]. A slightly inaccurate result can occur for the depth profiles but only for films with a high metal content when sputtering with a high incidence angle [357].

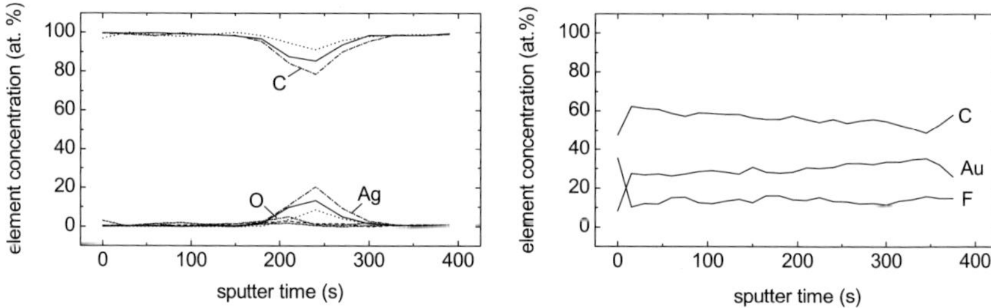

Fig. 3.15. XPS depth profiles for a plasma polymer multilayer system with embedded silver nanoparticles and continuously varying particle size for three different sample regions (*left*) and a plasma polymer gold composite film with continuous vertical particle distribution (*right*)

Figure 3.15 (left) shows the XPS depth profile for the chemical elements Ag, C, and O in a plasma polymer multilayer system containing silver nanoparticles (film I/9) for three different sample regions. The elements were detected using the binding energies at 284.5 eV (carbon C 1s), 531.0 eV (oxygen O 1s) and 368.3 eV (silver Ag $3d_{5/2}$). Photoelectrons from silver are first detected after a sputter time of $t_{sp} = 210$ s and have completely disappeared after $t_{sp} = 300$ s. Owing to the continuously varying metal content, the maximum proportions of silver atoms for the three sample regions are 20%, 13% and 8%, respectively, after $t_{sp} = 240$ s.

The depth profile of a multilayer system made from a first plasma polymer layer, a metal-containing plasma polymer layer, and a second plasma polymer layer shows that the metal particles are completely embedded into the polymer matrix over a large sample region (sputter area 4 mm^2, area analyzed about 1 mm^2). This is also shown in Fig. 3.16. It depicts the time behavior of the intensity of the Ag $3d_{3/2}$ binding energy at 374.3 eV and the Ag $3d_{5/2}$ binding energy at 368.3 eV for one of the three samples in Fig. 3.15. The left-hand picture of Fig. 3.16 gives a front view starting with the spectrum from the film surface. The first spectrum is the line near to the x–axis. In the first, second, third, fourth and fifth lines the peak intensity is zero. These lines correspond to sputter times $t_{sp} = 0$ s, 30 s, 60 s, 90 s, 120 s and 150 s. Silver is then detected, after a sputter time $t_{sp} = 180$ s. Maximum intensity is reached after $t_{sp} = 240$ s. The right-hand picture of Fig. 3.15 is a back view. Here, the spectrum taken after a sputter time 600 s is near to the x–axis. It shows the Ag $3d_{3/2}$ and Ag $3d_{5/2}$ binding energies starting with the spectrum at the substrate (interface between the plasma polymer and the silicon oxide substrate). This back view shows that there is also a plasma polymer film without silver particles between the substrate and the plasma polymer–silver composite film. Silver is only found in the sputter interval $180 < t_{sp} \leq 300$ s. The fact that embedding occurs slightly asymmetrically is interpreted as being due to the different sputter rates The weak shifts determined experimentally in the Ag $3d_{3/2}$ and Ag $3d_{5/2}$ binding energy peaks may result from incompletely compensated surface charge whilst the depth profile is being recorded.

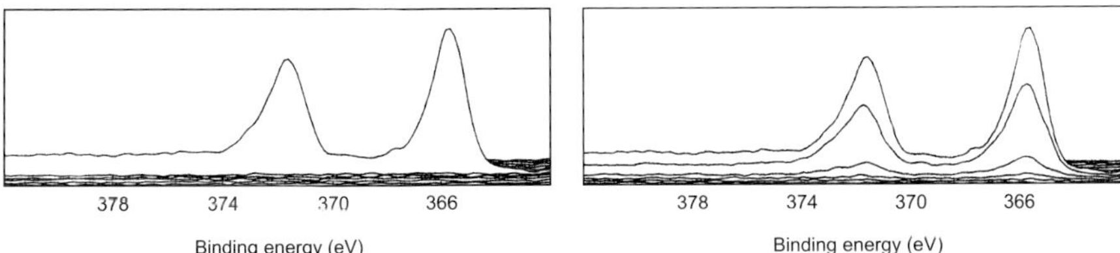

Fig. 3.16. XPS depth profiles for a plasma polymer multilayer system with embedded silver nanoparticles in a three-dimensional plot. In the *left-hand* sputtering profile, the curve near to the x–axis depicts the XPS spectrum on the film surface ($t_{sp} = 0$), and the upper curve is the XPS spectrum after $t_{sp} = 600$ s. In the *right-hand* sputtering profile the curve near to the x–axis depicts the XPS spectrum after $t_{sp} = 600$ s, and the upper curve is the XPS spectrum on the film surface

A sudden increase in the oxygen proportion (not represented in Fig. 3.15) shows when the substrate is reached, and takes place for all three samples after $t_{sp} = 660$ s. Combined with the fact that the silver particles were found after the same sputter times t_{sp} for all samples, this indicates that all sample from the multilayer have almost the same film thickness. The latter can be estimated using the experimentally determined etching rates for plasma polymer films without embedded nanoparticles (160 Å min^{-1} for a plasma polymer film made from benzene with $p = 0.85$ W cm^{-1}, Fig. 2.3). The effect of silver content on the total sputter time can be neglected, because the sputter rate for silver is significantly higher than for the plasma polymer, and the proportion of silver is not more than 20%. The film thickness $d_m \approx$

180 nm determined in this way corresponds to the film thickness $d_\mathrm{m} = 160$ nm which was determined by optical distance measurements.

Similar depth profiles to those shown in Fig. 3.15 (left) have been determined for multilayer systems with plasma polymer films containing gold, copper or indium particles [357]. All films with silver, gold or copper particles are free of oxygen after fabrication. Oxygen only adsorbs on the surface. The slight increase in oxygen level visible in Fig. 3.15 after $t_\mathrm{sp} = 210$ s results from increased spectrometer noise when silver is detected simultaneously. But when indium is detected for multilayer systems with indium–containing plasma polymer films, an oxygen proportion of up to 10% is simultaneously determined.

The right-hand picture of Fig. 3.15 shows the XPS depth profile of a plasma polymer–gold composite film made by plasma polymerization and metal sputtering (film III/1) and gives an example of the fabrication of a plasma polymer film with homogeneous vertical metal particle distribution. As shown in the depth profile, the gold content is almost constant at 25–30% for the total sputter time. The different elemental distribution occurring at the surface is caused by the high fluorine proportion there.

The depth profiles discussed show how the vertical embedding of metal particles can be investigated via XPS studies. Possible modification of multilayer systems, by a thermal treatment, for example, can be confirmed or ruled out by XPS depth profile investigation (Sect. 4.2.1).

3.2.3 Three-Dimensional Reconstruction of Particle Geometry

Since the sample preparation methods for lateral and vertical TEM micrographs cannot be used simultaneously, determination of the vertical and lateral size and shape is not feasible for one and the same metal particle. However, a three-dimensional reconstruction of the particle size and shape is possible by an appropriate choice of the samples.

Figure 3.17 shows lateral and vertical views of the silver particles embedded in a plasma polymer multilayer system (film I/7), each containing 110 silver particles, represented in the size and shape distribution histograms made by image analysis. The lower image of Fig. 3.17 depicts some of the silver particles, which became visible in the vertical cross section after a further thinning of the samples shown in Fig. 3.13. TEM micrographs were taken at various positions so that enough single particles were available for statistical analysis.

The mean particle size is $\bar{D} = 20.1 \pm 5.1$ nm for the lateral direction and $\bar{D} = 16.3 \pm 7.7$ nm for the vertical direction. Results for the shape factor give $\bar{S} = 0.89 \pm 0.04$ (lateral) and $\bar{S} = 0.83 \pm 0.06$ (vertical). However, when interpreting the result in the vertical direction, it must be remembered that the particles may be thinned over their maximum vertical principal axis, so that some particles appear smaller in the XTEM micrographs. This means that more small particles are observed in lateral TEM micrographs, with

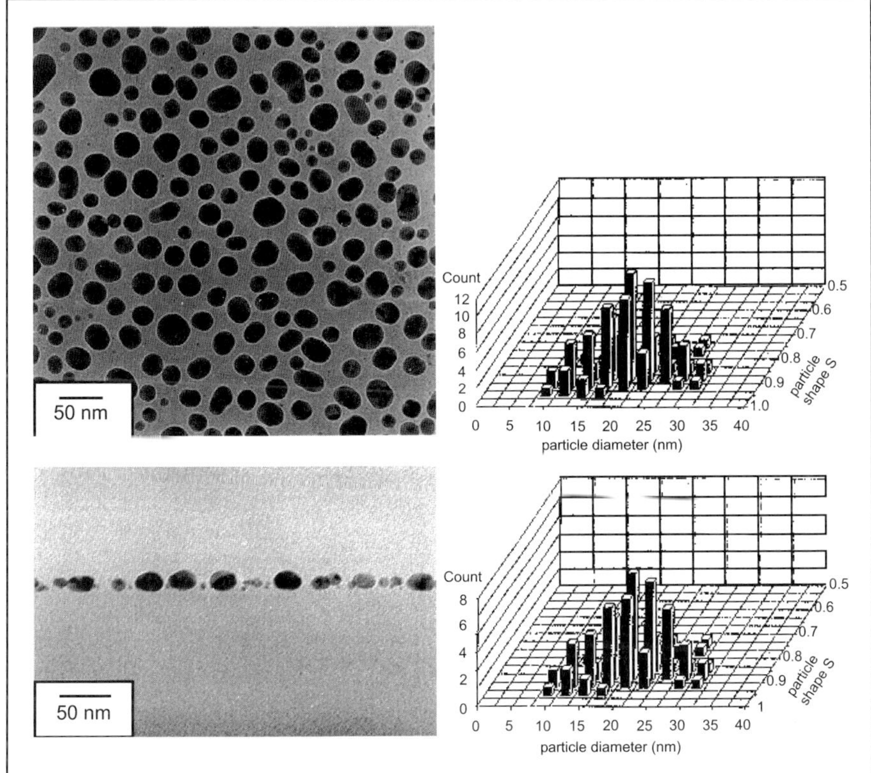

Fig. 3.17. Three-dimensional reconstruction of the nanostructure of a plasma polymer multilayer with embedded silver nanoparticles by means of TEM micrographs in the lateral and vertical directions

particle sizes of $1 < D \leq 10$ nm. The discrepancy between mean particle sizes in the vertical and lateral micrographs also arises through deviations from spherical particle shape. If the particles are assumed to be randomly oriented ellipsoids whose longest principal axis is parallel to the substrate, the longest principal axis is always the one seen in lateral TEM micrographs. This only works for the vertical micrograph if the longest principal axis is randomly parallel to the cross-sectional plane.

On the basis of results from lateral and vertical TEM micrographs, embedded silver particles can be simply described as spheroids or ellipsoids with a principal axis parallel to the substrate. This assumption also applies to embedded gold particles.

3.3 Surfaces and Intermediate Layers

3.3.1 Crystal Structure of Embedded Particles

As already mentioned in Sect. 3.2.1, larger embedded metal particles are polycrystalline. Figure 3.18 presents bright-field and dark-field images of the same sample position from a plasma polymer film with embedded silver particles (film I/1). The polycrystallinity of the particles is clearly visible. Different orientations of crystal planes inside particles show up as different contrasts in the TEM micrographs. Particles in Bragg position give the best contrast in the bright-field image and appear white in the dark-field image.

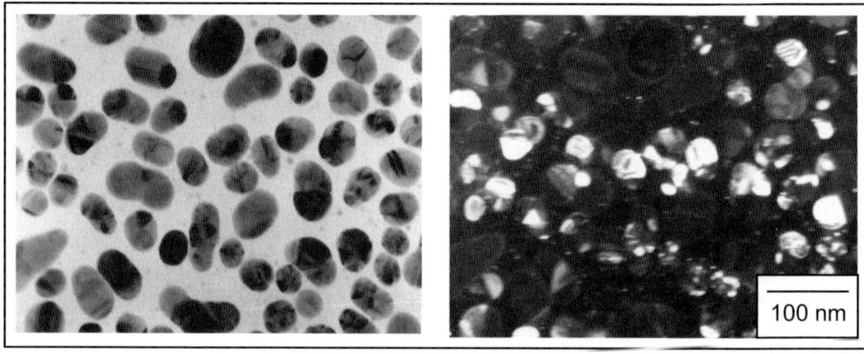

Fig. 3.18. TEM micrographs showing silver particles embedded in a plasma polymer film. *Left*: bright image, *right*: dark image of the same sample position

High resolution TEM micrographs (HRTEM) provide further information about the crystal structure of embedded metal particles. Particles with sizes $D < 20$ nm can be completely imaged at high resolution with single micrograph. Individual particles can be investigated on lateral samples as well as cross-sectionally prepared samples, under conditions where good contrast is achieved for the high resolution electron microscope image.

Figure 3.19 shows a high resolution XTEM micrograph of a silver particle embedded in a plasma polymer matrix (film I/7). Most smaller particles embedded in a plasma polymer film show an undistorted face-centered cubic (fcc) lattice. The silver particle shown in Fig. 3.19 has a decahedral structure (multiply twinned particle), as is also found, for example, for much smaller silver particles ($D \approx 5$ nm) on carbon films [358] and gold particles ($D \approx 3$ nm) on SiO_2 or MgO [359, 360] as well as e.g. at polymer coated gold particle ($D = 15$ nm) [361], but also particles without crystal defects were found, for example, for silver particles embedded in glass [362] crystallites without Changes in crystal structure have been detected during electron irradiation. Single crystalline gold particles ($D \approx 2$ nm) change to multiply twinned particles and conversely [360, 363]. Depending on the number

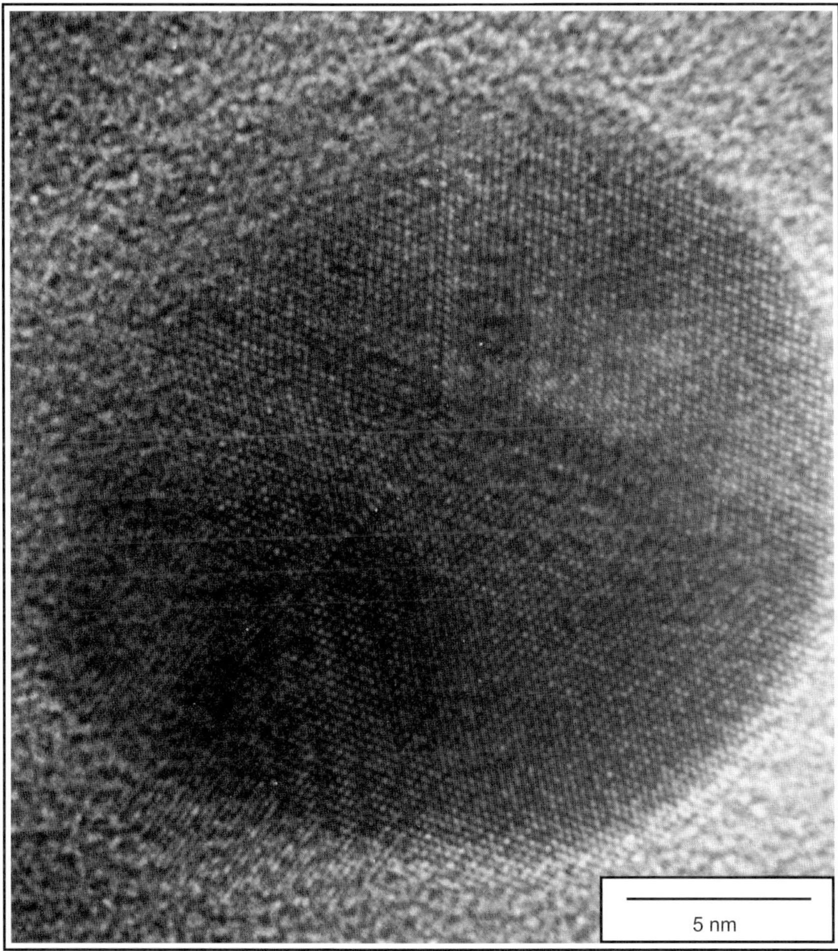

Fig. 3.19. XTEM micrograph of a silver particle embedded in a plasma polymer film

of atoms ($N \leq 1\,000$), both crystal structures represent a minimum energy state of the particle [364]. In addition, rotation of gold particles has been observed during electron irradiation in HRTEM [365]. For larger silver particles ($D \geq 10$ nm) made by evaporation in an inert gas atmosphere, decahedral crystallites were found [366, 367].

Larger embedded indium particles (Fig. 3.20) do not show multiply twinned particle growth. In addition, very small indium particles possess an almost undistorted tetragonal lattice (see Fig. 4.24). Figure 3.20 shows two HRTEM micrographs of the interface between indium particles and plasma polymer for a multilayer system containing indium nanoparticles (film II/1).

The lattice plane spacing was determined to be 2.76 Å, which deviates by less than 3% from the indium lattice plane spacing in the (101) direction [368].

The crystal structure can be determined with atomic resolution by high resolution electron microscopy in spite of the plasma polymer layer surrounding the particle. However, this is of only secondary importance for correlating the electrical and optical properties of the metal particles.

3.3.2 Intermediate Layers

There is so far no evidence for a special interface reaction between the different embedded metal particles and the plasma polymer. It is quite possible that a modified polymer structure forms on the many individual interfaces between the plasma polymer and metal particles.

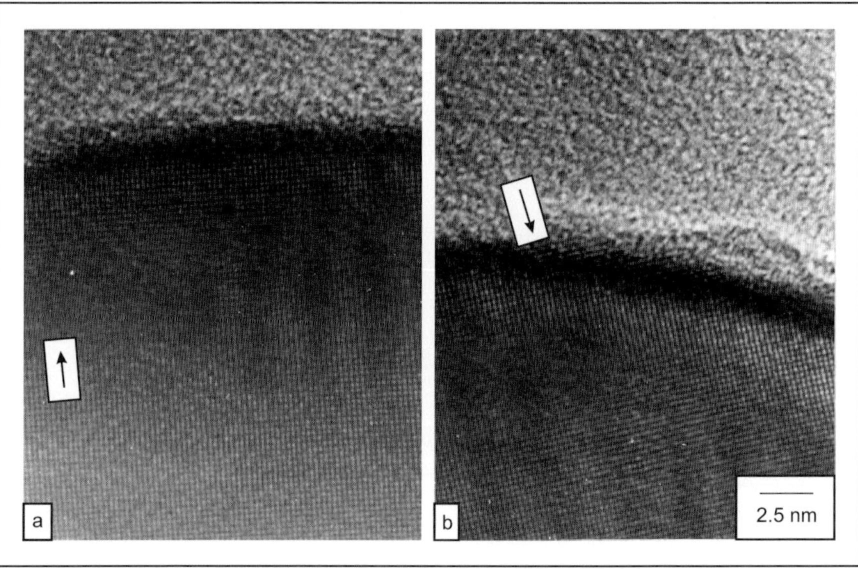

Fig. 3.20. (a,b) HRTEM micrograph of the indium particle/plasma polymer interface

Figure 3.20 shows a plasma polymer/metal particle interface at high resolution. No crystalline structures are visible in this region. The lattice planes are sharply imaged up to the edge of the particle (see arrow in Fig. 3.20). The encapsulation is similar to TEM micrographs of gold particles with stabilized ligand, formed by the fabrication of gold particles from a colloid solution [369].

The indium particles shown in Fig. 3.7c on page 55 exhibit concentric encapsulation. These encapsulation formations can be explained by a strengthened diffraction contrast of the plasma polymer matrix. A similar encapsulation formation has also been observed for embedded silver particles, but after thermal treatment (see Fig. 4.5d). One reason for the strengthened contrast around the metal particle could be a chemically modified plasma polymer matrix close to the metal particles. Alternatively, very small silver clusters or silver atoms could be responsible for the contrast amplification. However, there is no further evidence for either hypothesis.

In previous descriptions the plasma polymer has been discussed as a laterally homogeneous material with a specific surface morphology. Figure 3.21 shows that interfaces can form within the plasma polymer.

Figure 3.21 shows an XTEM micrograph with the whole vertical film setup for a multilayer system consisting of three plasma polymer layers and two plasma polymer layers with embedded copper particles (film I/13). The lower pictures show zoomed regions of the metal-containing layers. The outer plasma polymerization conditions (power density, monomer flow rate) were constant. Copper was evaporated twice, after $t_{pp} = 2$ min and after $t_{pp} = 12$ min, in order to investigate the influence of an embedded metal on the morphology of the plasma polymer. Clear changes occur 50 nm above the embedded copper nanoparticles in the plasma polymer film. As shown in the detailed zoom images, the plasma polymer layer, which grows initially above the copper–containing layer without a visible morphology, becomes more structured after reaching a film thickness of 50 nm.

This modification of the plasma polymer morphology during film growth can be explained by the changed deposition conditions caused by embedded copper particles. The aluminum foil of the metal electrodes in the plasma reactor are covered by the plasma-polymerized layer. This causes changes in polymerization conditions, as confirmed by optical emission spectroscopy (OES) during the time progression of the plasma discharge [40]. The result is a change in plasma polymer morphology. After copper particle evaporation, the plasma polymer grows again on an approximately electrically conductive substrate. When the copper particles are completely covered with the growing layer, conditions for plasma discharge revert to their previous state.

Figure 3.21 shows that local interfaces also occur in the plasma polymer among the intermediate layers between particles and plasma polymer. These inner plasma polymer interfaces may be important in modifying the nanostructure.

Figure 3.21 also shows an intermediate layer of the plasma polymer and the SiO_2 substrate as in Figs. 3.13 and 3.14. XTEM investigations do not reveal any special features on this interface and XPS analysis has not provided any differing information.

3.3.3 Film Surface

The surfaces of plasma polymer multilayer systems containing metal nanoparticles are not distinguishable from surfaces of plasma polymers without metal embedding made under the same fabrication conditions, as long as the embedded metal particles are small compared to the film thickness of the sur-

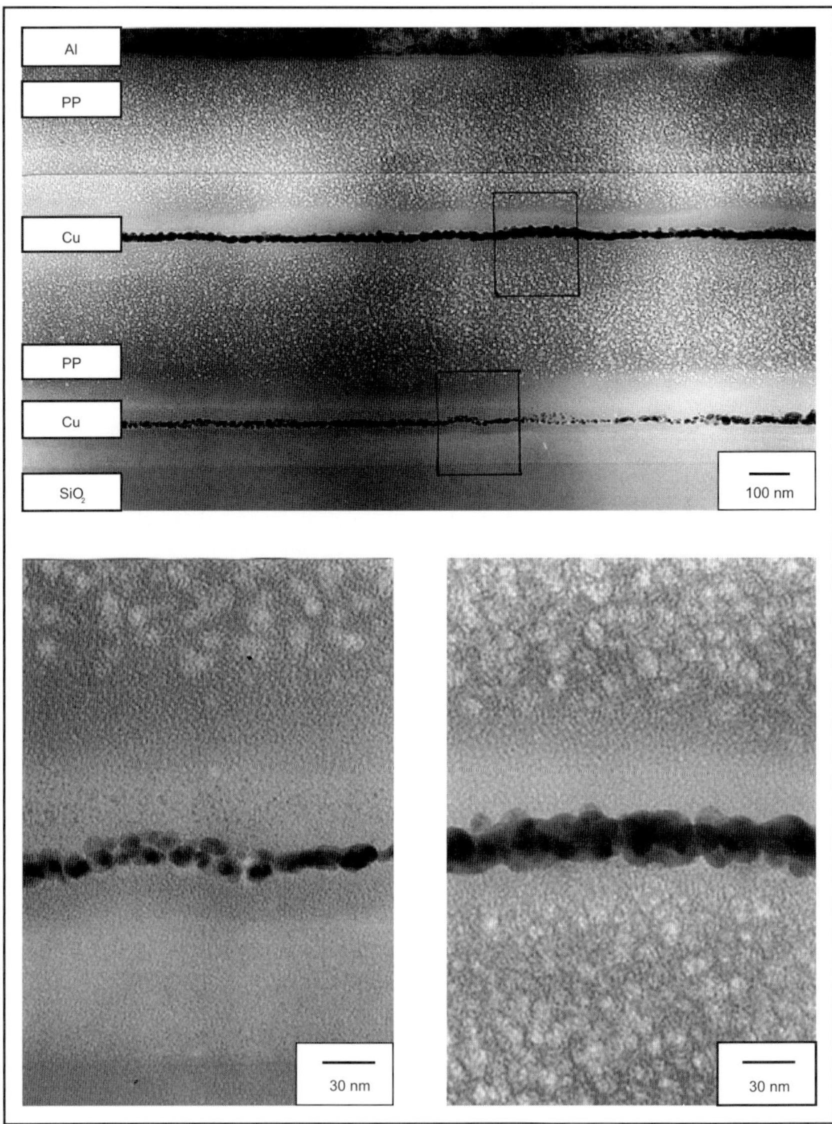

Fig. 3.21. XTEM micrographs (total multilayer system and zoomed areas) for a plasma polymer multilayer system with embedded copper particles

Fig. 3.22. SEM micrographs of a plasma polymer multilayer system with embedded indium nanoparticles

rounding plasma polymer. Atomic force microscopy (AFM) investigations have confirmed this.

For larger metal particles ($D \gg 100$ nm) the surface of the composite film is equivalent to the surface morphology of the metal particles, because the thin plasma polymer layer cannot compensate for the surface roughness. Accordingly, scanning electron microscope (SEM) micrographs show a hill structure for the plasma polymer–metal composite films with higher metal contents, which is also typical for thin metal films deposited at room temperature. Figure 3.22 shows the surface of a plasma polymer multilayer system (film I/11) containing indium nanoparticles. Similarly formed surface morphologies have been observed on plasma polymer films containing germanium or tin.

Since the particles are completely embedded for multilayer systems with metal particles having $D \leq 10$ nm, it can be assumed for further investigations of nanostructural modifications that the film surface and plasma polymer/substrate interface are of secondary importance.

4. Nanostructural Changes

4.1 Nanostructural Changes in Embedded Nanoparticles

4.1.1 Overview

Structural changes affecting nanoparticles embedded in an insulating matrix material are generally defined as changes in the size and shape distributions of the embedded particles. This includes changes in the inner and outer interfaces, or the chemical structure of the matrix material. The simplest way to bring about nanostructural change is thermal treatment of the films.

The possibilities for and progression of nanostructural changes are mainly determined by the matrix material (insulator). If there is no matrix, as in the case of thin discontinuous metal films, diffusion of metal atoms or metal particles is possible along the substrate. Long-term stability is determined mainly by properties of the interface between metal particles and substrate, and thermal treatment accelerates surface diffusion, e.g., for gold particles on KBr [370], SiO_2 [371] or polystyrol [372].

The size and shape distributions of metal particles in a metal oxide film (Al_2O_3, MgO) or in glass show good long-term stability, but thermal treatment from several hours up to a few days leads to significant changes in particle size and shape distributions, e.g., for silver particles in glass at 870 K [373]

Metal nanoparticles embedded in a polymer matrix also show adequate long-term stability as regards size and shape distributions. However, changes caused by thermal treatment take place at lower temperatures. As an example, clear changes occur for the particle size and shape distributions in PTFE films containing gold nanoparticles treated at temperatures of 455 K for 60 min [256]. On the basis of these results, changes were observed in situ during thermal treatments at lower temperatures using the electron microscope.

For further discussion, various types of material transport in thin films must be distinguished. These mechanisms for nanostructural change are caused by atomic diffusion through the matrix (Ostwald ripening), atomic diffusion along the particle surface (recrystallization and coalescence) and particle migration (Fig. 4.1). All these processes are considered under the assumption of a closed system, in other words, supposing that no material is lost or supplied during nanostructural changes.

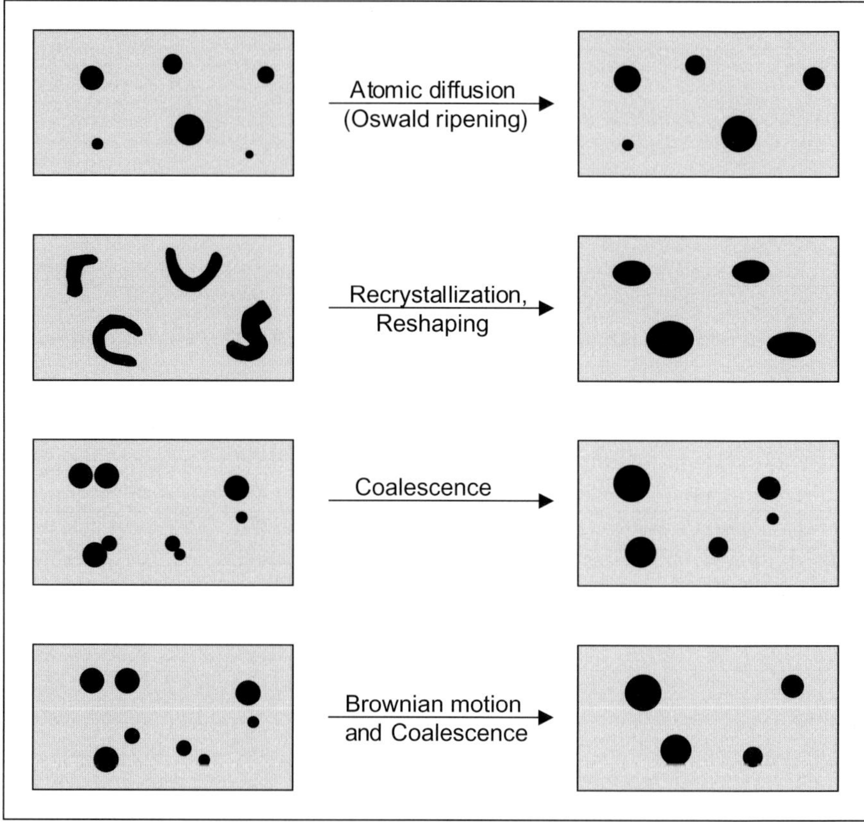

Fig. 4.1. Nanostructural changes in metal-particle-containing plasma polymer films

The proportion of surface energy is much increased for small particles, compared to the total binding energy, due to the high proportion of surface atoms. Thermal activation by annealing, laser irradiation or electron irradiation leads to increased atomic mobility. Because of the different surface energies, atomic diffusion from smaller to larger particles occurs through the matrix after an extensive thermal treatment. This process for size and shape changes in embedded particles is called Ostwald ripening. The so-called LSW theory describes this process for long annealing times and small filling factors, and also gives information about particle size statistics (see Sect. 4.1.2).

A second process is based on minimization of the surface energy of non-spherical particles (recrystallization) or pairs of contacting particles (coalescence). Both lead to the formation of more spherical particles by the diffusion of metal atoms along grain boundaries and along the particle surface. The term recrystallization is used if one non-spherical particle changes into a more spherical particle. Coalescence is used if two or more particles coalesce into one particle.

Further processes involve particle mobility in the matrix, which can take place if the matrix material is heated close to melting point, or if a polymer is above the glass transition temperature after thermal annealing. If two particles collide because of their movement in the matrix, this can also lead to coalescence.

The next section describes processes which create nanostructural changes in insulator films with embedded nanoparticles. In addition, possible chemical changes in the films and the decrease in melting point of very small particles are discussed. After these considerations, experiments to study diffusion processes are presented. Different sources of thermal energy input (thermal treatment, laser irradiation, electron irradiation) were used to realize various experimental conditions to observe different diffusion processes. The main objective is to perform in situ experiments in the transmission electron microscope.

Other methods for studying metal diffusion in polymers are radiotracer and Rutherford backscattering measurements. The diffusion of metals in polymers with the focus on chemical metal polymer interactions is reviewed in [374, 375].

4.1.2 Atomic Diffusion and Ostwald Ripening

The enlargement (coarsening) of particles embedded in a solid or liquid matrix is known as Ostwald ripening[1]. It takes place by atomic diffusion from smaller to larger particles during thermal treatment close to the melting point as a consequence of the different surface energies of particles of different sizes. The theoretical basics were established in 1961 by Lifshitz and Slyozov [377] and Wagner [378] (LSW theory).

Based on the Gibbs–Thomson equation, the solution concentration $c(R)$ around a particle with radius R is

$$c(R) = c_\infty \exp\left[\frac{2\gamma\Omega}{k_{\mathrm{B}}TR}\right] , \tag{4.1}$$

where c_∞ is the solution concentration in equilibrium for the particle/matrix interface, γ the specific interface energy between particle and matrix, Ω the atomic volume of the particle material, k_{B} the Boltzmann constant and T the absolute temperature. For particles with $R \geq 100$ nm [379] and small interface energies, the equation can be linearized to give

$$c(R) = c_\infty \left[1 + \frac{2\gamma\Omega}{k_{\mathrm{B}}TR}\right] \tag{4.2}$$

Consequently, the solution concentration $c(R)$ for a particle with radius R is always $c(R) \geq c_\infty$, and $c(R_1) > c(R_2)$ for two particles R_1, R_2 with $R_1 < R_2$.

---

[1] W. Ostwald [376] first discovered the increased solubility of mercury oxide particles in 1900.

The result is an atomic diffusion from smaller to larger particles. Applying Fick's law of diffusion (4.3) for the diffusion flow,

$$j = -D_\mathrm{d} \left[\frac{\delta c}{\delta R} \right] , \tag{4.3}$$

where D_d is the diffusion coefficient. A growth law can then be determined for the precipitation whereby the mean particle radius satisfies (LSW theory)

$$\bar{R}^3(t) - \bar{R}_0^3(t_0) = K_\mathrm{LSW}\, t . \tag{4.4}$$

In this case, $\bar{R}_0^3(t_0)$ is the mean particle radius at the start t_0 of the thermal annealing process, $\bar{R}^3(t)$ is the mean particle radius at time t and K_LSW is a proportionality constant defined by

$$K_\mathrm{LSW} = \frac{8}{9} \frac{\sigma \Omega^2 c_\infty D}{k_\mathrm{B} T} . \tag{4.5}$$

From the so-called rate equation (4.4), one can see that the radii of embedded nearly spheroidal particles are proportional to the cube root of the annealing time. After long thermal annealing (asymptotic limiting case) the relative particle size $p = R/\bar{R}$ satisfies a time independent distribution function $h(p)$ given by

$$h(p) = p^2 \left(\frac{3}{3+p} \right)^{7/3} \left(\frac{3}{3-2p} \right)^{11/3} \exp\left(\frac{-2p}{3-2p} \right) . \tag{4.6}$$

The distribution function is highly asymmetric (see Fig. 4.2). Particles larger than $3R/2$ are not permitted $h(p \geq 3/2) = 0$, and the following normalization condition holds:

$$\int_0^\infty h(p)\mathrm{d}p = 1 . \tag{4.7}$$

A version of LSW theory has been derived by Ardell [380] and applied to two-dimensional diffusion processes. The rate equation (4.4) is still used, but with the modified distribution function

$$h(p) = p^2 \left(\frac{2}{3} \right)^3 \left(\frac{3}{3+p} \right)^{17/9} \left(\frac{3}{3-2p} \right)^{28/9} \exp\left[\frac{2}{3} \left(\frac{-2p}{3-2p} \right) \right] , \tag{4.8}$$

illustrated in Fig. 4.2. The cutoff $h(p \geq 3/2) = 0$ still applies.

Along with the restrictive assumption of infinitely small filling factors ($f \approx 0$), there are a number of other effects which influence nanostructural changes in a real mixed system and are not considered in the LSW theory. The coalescence of two contacting particles, also known as short diffusion, and possible particle migration, belong to this category.

A proportionality constant $K(f)$ and related function $g(f)$, both depending on the filling factor, are introduced to extend LSW theory to significantly larger filling factors ($f > 0$):

$$\bar{R}^3(t) - \bar{R}_0^3(t_0) = K(f)t , \quad \text{where} \quad K(f) = g(f)K_\mathrm{LSW} . \tag{4.9}$$

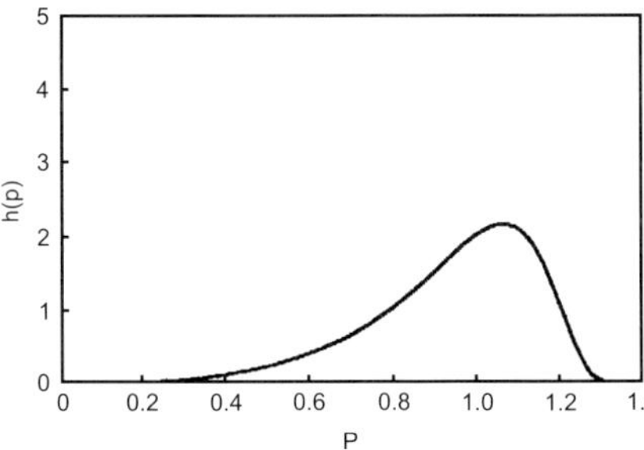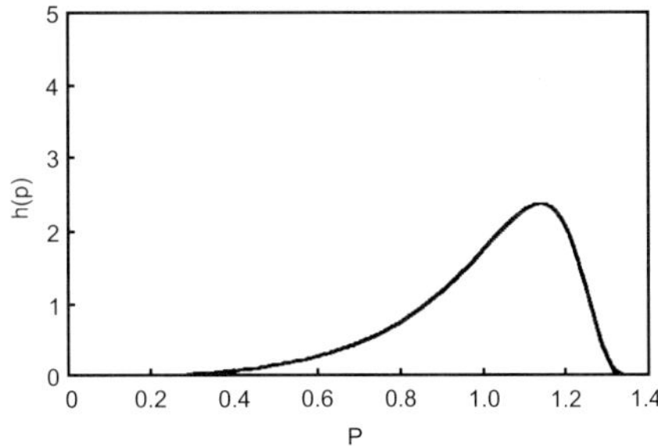

Fig. 4.2. Time independent particle distribution functions $h(p)$ according to LSW theory. *Left*: three-dimensional diffusion governed by (4.6). *Right*: Two-dimensional diffusion governed by (4.8)

There are a number of theories for the calculation of distribution functions which depend on the filling factor, sometimes mutually contradictory. These theories suggest different nonlinear functions $g(f)$ which increase rapidly with the filling factor [381–384]. Distribution functions for the particle sizes become wider and more symmetric with increasing filling factor and differ greatly from the LSW distribution (4.6).

A number of distribution functions $h(p)$ were developed by Voorhees and Glicksman [381] for different filling factors under the assumption of many-particle diffusion, taking into account diffusion interactions between the particles. The size distribution during the beginning of thermal treatment depends strongly on the initial distribution. Size distributions based on the work of Voorhees and Glicksman (Fig. 4.3) have been determined from the resulting particle histograms under these assumptions and, at the time of writing, are only available in numerical form. With these distribution functions, particles with $p \geq 3/2$ are permitted. Particle distribution functions are still subject to the normalization conditions (4.7).

The modified LSW theory due to Davies, Nash and Stevens [383] considers the case where direct contact can occur between neighboring particles during their growth. Particle distances remain constant. For increasing filling factors, the particle distribution functions (Fig. 4.3) derived from this approach grow wider and more symmetric. Particles larger than $p \geq 3/2$ are allowed. Further size distributions with filling factor dependence have been proposed by Brailsford and Wynblatt [382]. Here, particles with $p \geq 3/2$ are not allowed. Otherwise it is possible to describe the particle growth process using numerical simulations based on the Cahn–Hilliard equation [385].

LSW theory is applied in order to understand experimental results for particle enlargements in metallic particle–matrix systems, consisting of solid–solid dispersions (Ag–Ni alloys [386]), solid–liquid dispersions (liquid phase sintering of Co–Cu alloys [387,388], Cu–Pb alloys [389], Fe–Cu alloys [388]), liquid–solid dispersions (Ag–Ni alloys [386]) and liquid–liquid dispersions

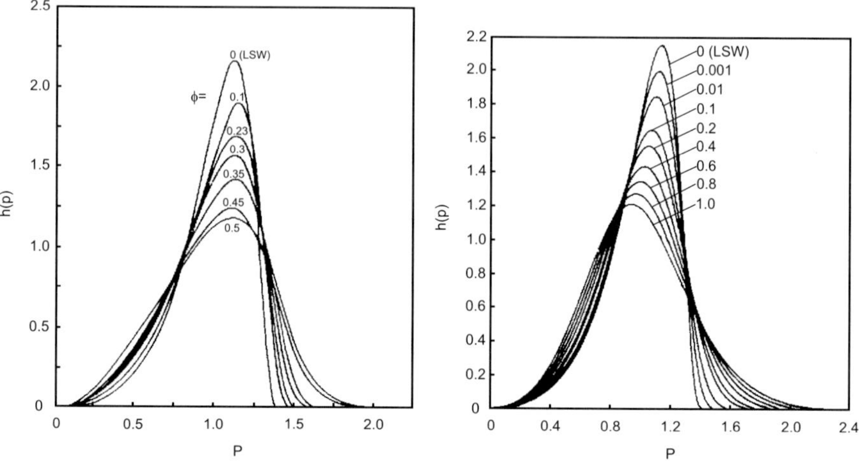

Fig. 4.3. Time dependent particle distribution functions for higher filling factors, after Voorhees and Glicksman [381] (*left*) and Davies and Nash [383] (*right*)

(Pb–Zn alloys, Pb–Al alloys [390]). In these cases, the rate equation (4.4) generally holds. LSW theory is also applied for metal silicide films during thermal treatment of ion-implanted silicon [391, 392]. Furthermore, LSW theory is used to describe size changes in polymer mixtures (polybutadiene particles in polyethylene [393]) or for the growth of gas bubbles in metals (10 nm helium bubbles in precious steel [394]).

Ostwald ripening has also been investigated for tempering of silver particles [373, 395] or gold particles [514] in a glass matrix. In [395], the glass with embedded silver particles (particle sizes from 100 nm to μm) was annealed up to 873 K, almost the melting point (881 K) of the glass in question. The particle size distribution was then determined after long time periods of up to 168 h annealing time. After longer annealing times, a particle size distribution was found that was at least similar to LSW predictions (4.6). The rate equation (4.4) was thereby confirmed. Different rate constants K_{LSW} were calculated for different temperatures [373], and the interface energy of $\gamma_{\mathrm{Ag}} = 1.0\ \mathrm{J\,m^{-2}}$ was determined for the atomic diffusion of silver and gold at 870 K [373].

For application of Ostwald ripening to thin polymer with embedded nanoparticles, it must be shown that interparticle atomic diffusion through the matrix material is the dominating factor. This is not the case for thin, discontinuous metal films supported by various substrate materials, because the very high atomic surface mobility is determined by interface properties between substrate and particle and not decreased by any matrix material. For this reason, no detailed description is known using LSW theory for nanostructural changes in thin discontinuous metal films. Only in [351] has Ostwald ripening been considered as a possible process for changes in nanostructure.

In order to detect atomic diffusion in plasma polymer films with embedded nanoparticles, long thermal treatment times at relatively low temperatures are necessary. This is because the thermal treatment has to be carried out below the temperature at which pyrolysis of the polymer matrix occurs.

4.1.3 Coalescence and Recrystallization

The term coalescence is used in different areas of physics, where single isolated atoms, clusters, particles or drops grow together to form larger clusters, particles or drops. A distinction must be made between dynamic and static coalescence. During dynamic coalescence, the mass of the system increases, whilst it remains nearly constant during static coalescence.

If the coalescing medium is not closed, i.e., extra atoms, clusters or particles are added to the system, the process is called coalescence for an open system or dynamic coalescence. Dynamic coalescence is a growth process for thin films and occurs, for example, if single metal particles grow together during metallic film deposition from the gaseous phase.

The so-called static coalescence in granular materials takes place after fabrication of the particles, via an additional thermal treatment or an aging process or as a result of some mechanical energy supply. No material is added during the coalescence process. Of course, if static coalescence has started it has its own dynamic. An example of static coalescence is provided by evaporated gold particles growing together during thermal treatment [371]. In the present case of polymer thin films with embedded nanoparticles, only nanostructural changes in films already deposited are discussed. Consequently, only static coalescence will be discussed.

The changes in shape and size of embedded metal particles taking place at temperatures far below the melting point of the metal, by diffusion of metal atoms along the particle surface or along inner grain boundaries, are designated as coalescence and recrystallization. If the number of particles remains approximately constant or only one particle is under consideration, the term recrystallization is used. All diffusion processes causing a change in shape of one of the particles can also be called reshaping. Coalescence is used if two or more particles join together to form one particle during the diffusion process. The number of particles then decreases. In principle, coalescence and recrystallization result from the same diffusion processes which will now be described.

First consider the change in two contacting particles during coalescence, as shown in Fig. 4.4. A neck radius b is determined by means of the concave contact area of two particles with radius R. The contact area between the particles coming together is caused by atomic diffusion along the surfaces of both particles. The concentration of the solution $c(b)$ on the contact area is given by [391, 396]

$$c(b) = c_\infty \left[1 - \frac{\gamma \Omega}{k_\mathrm{B} T b} \right] , \qquad (4.10)$$

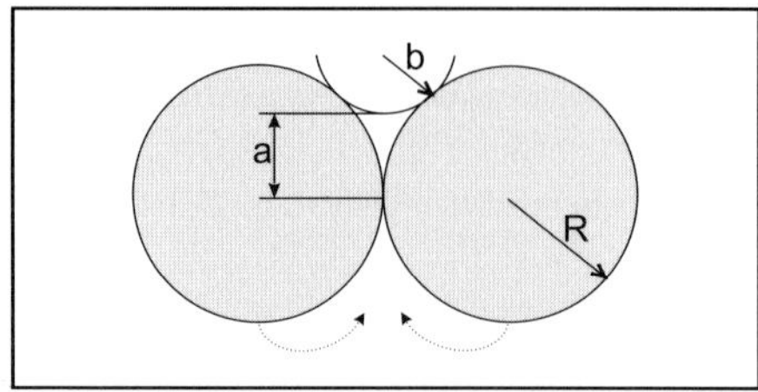

Fig. 4.4. Coalescence of two metal particles

where c_∞ is the solution concentration in equilibrium for the particle/matrix interface, γ the specific interface energy between particle and matrix and Ω the atomic volume of the particle material. This formula is similar to the Gibbs–Thomson equation (4.2). However, the solution concentration on the contact area is smaller than the solution concentration in equilibrium. The result is an atomic diffusion of metal atoms on the surface of the particles toward the contact area of both particles.

Growth rates can be determined for the closing together of the contact areas of two particles from geometric derivations. Using Fick's diffusion law (4.3) [397],

$$\text{volume diffusion} \qquad a^5 = \frac{20\gamma\delta^3 R^2 D_\mathrm{d}}{k_\mathrm{B}T}t\,,$$

$$\text{surface diffusion} \qquad a^7 = \frac{28\gamma\delta^4 R^3 D_\mathrm{d}}{k_\mathrm{B}T}t\,, \qquad (4.11)$$

$$\text{grain boundary diffusion} \qquad a^6 = \frac{960\gamma\delta^4 R^2 D_\mathrm{d}}{k_\mathrm{B}T}t\,,$$

where δ is the atomic radius and D_d the diffusion coefficient. These equations are derived under the simplified assumption that the lattice oscillation and binding energy between two atoms do not change for different curved surfaces. The time needed for the interfaces between two particles to disappear (sintering) can be estimated with (4.11). An estimation of this type has been carried out [397] for silver particles deposited in an inert gas atmosphere. In this case, a temperature-dependent diffusion constant D_d with an activation energy of surface diffusion W is assumed, viz.,

$$D_\mathrm{d} = D_0\exp\left[-\frac{W}{k_\mathrm{B}T}\right]\,. \qquad (4.12)$$

For the calculations, a self-diffusion of silver with $D_0 = 0.9$ cm^2 s^{-1} and a surface energy of $\gamma_\mathrm{Ag} = 2$ J m^{-2} were assumed. Following [397], the coalescence time t_co is the time for two particles ($R = 10$ nm) to form an interface of $a = 1$ nm at different temperatures (see Table 4.1).

When using these coalescence times at different temperatures for experimental results, it should be remembered that the diffusion constants D_0 for silver embedded in other materials are smaller than for self-diffusion (e.g.,

Table 4.1. Coalescence time for two silver particles with $R = 10$ nm

Temperature T [K]	Volume diffusion t_{co}	Grain boundary diffusion t_{co}
298		5 years
373	15 years	25 days
473	21 min	1 days
573	0.3 s	4 min
673	0.9 ms	20 s

$D_0 = 0.03$ cm^2 s^{-1} for Ag in Cu [397]), and that coalescence times are strongly influenced by these lower diffusion constants. They are also affected by the generally unknown values for the specific interface energy between particle and matrix.

Estimating coalescence times for different types of diffusion, it is found that coalescence progresses very fast at temperatures over ≈ 550 K. For this reason the process is also called liquid-like coalescence [351, 398]. Estimated coalescence times show that coalescence is in principle possible at room temperature, although at a very slow rate.

According to similar estimates, coalescence times of $t_{co} = 10^{-6}$ s were determined for the coalescence of gold particles ($R = 25$ nm) embedded in plasma polymer films at 450 K [256]. The coalescence time for gold particles with $R = 5$ nm has been estimated as $t_{co} = 10^{-8}$ s at $T = 673$ K [399].

In situ investigation of coalescence by electron microscopy is no simple matter due to the very fast rate of coalescence at higher temperatures. The growing together of gold particles on amorphous carbon and the changing interface between two gold particles have been investigated using high resolution electron microscopy at 293 K, and atomic diffusion along the surface has been detected [400]. Coalescence was stimulated by electron irradiation (300 keV, 2 nA µm^{-2}) in the electron microscope and by the resulting heating.

These briefly expressed ideas about the coalescence of metal particles can be used for a phenomenological description of nanostructural changes in plasma polymer films with embedded nanoparticles. Predictions for particles before or during coalescence are not possible to present. The influence of the metal particle-encapsulating matrix on coalescence is not considered.

4.1.4 Migration of Embedded Metal Particles

In previous considerations it was assumed that metal particles are solidly embedded in the surrounding matrix and do not change position. But if metal particles are embedded in a polymer matrix, the lower thermal stability of the polymer matrix, e.g. compared to a metal oxide matrix, must be considered.

During thermal treatment, a conversion from a plastic into a soft elastic state takes place at the glass transition temperature of a polymer. Close to the glass temperature, polymer chain segments start to move under the effects of supplied heat (micro-Brownian movement), although they remain bonded onto individual traps. The polymer becomes elastic. If the polymer is heated above the glass temperature up to the yield temperature, bonds break from the traps and freely moving macromolecules can exist (macro-Brownian movement).

This model is very difficult to use for plasma polymer films because of the non-uniform chain length and crosslinks between chains, and also because of the special adhesion between substrate and plasma polymer. It can be assumed that bulk values for the glass transition temperature of different polymers are not applicable to plasma polymer films. Furthermore, due to adhesion between polymer and substrate, it can be assumed that at high temperatures micro- and macro-Brownian movement occur simultaneously in plasma polymer films. However, embedded metal particles may be able to change location by possible movements of molecular chains.

Detailed experimental and theoretical methods for determining the glass temperature of plasma polymer films are not known. A glass transition temperature of 435 K is assumed for plasma polymer films made from fluorine hydrocarbons (C_3F_8) [258]. As shown in [343], when different metals were evaporated on a heated polymer substrate (styrene/hexamethylacrylate copolymer at 390 K), an embedding of silver or indium particles could be detected beneath the polymer surface. This effect has also been discussed in terms of the Brownian motion of metal particles in a heated polymer, which is similar to the migration of gold particles in polystyrol during thermal treatment (5 days, 450 K [372]).

When discussing experimental observations of nanostructural changes affecting embedded nanoparticles in plasma polymer thin films, it must be recognized that they are strongly influenced by the glass transition and melting temperatures.

4.1.5 Chemical Changes and Particle Oxidation

As a result of their higher interface energy, metal particles have greater chemical activity and enhanced catalytic properties compared with the bulk form [401]. It is conceivable that the presence of metal atoms or particles during plasma polymerization might lead to changes in the chemical structure of plasma polymer films as they develop. However, no studies have yet been made concerning either the influence that the catalytic properties of different metal particles might have on the chemical structure of plasma polymer films or of nanostructural changes in nanoparticles embedded in such a matrix material.

If an oxygenated monomer such as HMDSO is used, the metal may also partially oxidize during film deposition, and this oxidation may slowly continue after film fabrication. Furthermore, in a plasma polymer matrix, oxygen may diffuse through the plasma polymer matrix to the embedded metal particles, leading to a gradual oxidation of the particles.

Assuming that there is no initial oxide layer, and that all volume changes are negligible, a simple rate law can be defined for the growth of the oxide on the metal particle [402]:

$$t_{\mathrm{ox}} \sim \frac{R_{\mathrm{ox}}}{2} \left[1 - \frac{2}{3} \frac{R_{\mathrm{ox}}}{R_0} \right] , \tag{4.13}$$

where t_{ox} is the oxidation time, R_{ox} the thickness of oxide encapsulation and R_0 the particle radius before oxidation. The radius of the remaining inner metal particle is defined as R_{co}, with $R_{\mathrm{co}} + R_{\mathrm{ox}} = R$. From (4.13), oxidation clearly progresses faster for smaller particles. For large particles, $R_{\mathrm{ox}}/R_0 \ll 1$ and there is an approximate dependence $t_{\mathrm{ox}} \sim R_{\mathrm{ox}}/2$ between oxidation time and thickness of the oxide coating.

In [403] it is assumed that oxidation is incomplete, and that a particle core remains that is not oxidized. There is a linear dependence between the total radius R and the remaining metal particle core R_{co}. The ratio R/R_{co} is constant. This linear dependence is confirmed by experimental data for the oxidation of silicon particles at high temperature [404] or metal particles (Al, Mg, Be or Sn) at room temperature obtained by high resolution electron microscopy [405]. The formation of an oxide coating on silicon particles has been investigated by IR spectroscopy [406]. A linear growth behavior has been observed for the surface oxidation of thin copper films by ellipsometric investigations, whereas a volume growth of the oxidized metal particle is assumed [407].

A nanostructural change caused by oxidation of embedded metal particles can be detected from the electron diffraction pattern in transmission electron microscopy. While oxidation of gold and silver particles can be neglected, oxidation of copper and indium particles can take place during or after deposition. These thin oxide films consisting of only a few monolayers are only visible in high resolution TEM micrographs. The oxidation of copper particles can be detected, for example, by changes in the plasma resonance absorption of the embedded copper nanoparticles.

4.1.6 Melting Point Depression in Nanoparticles

Melting point depression for very small particles was described for the first time in [408], where an electron diffraction camera was used. The melting temperature is defined as the point at which Debye Scherrer rings start to change from sharp to diffuse. A detailed study of melting point depression can be found in [13].

Upper and lower bounds for melting point reductions can be estimated from thermodynamic considerations [13]. The lower bound can be determined by considering the free energy of particles in the liquid and solid states.

Following [13], the assumption of immediate melting after the particle is formed from a melt leads to a minimum value for the decreased melting temperature $T_{\mathrm{R}}^{\mathrm{min}}$ given by

$$T_{\mathrm{R}}^{\mathrm{min}} = T_0 \left[1 - \frac{1}{w_0 R} \left(\frac{\gamma_{\mathrm{s}}}{\rho_{\mathrm{s}}} - \frac{\gamma_{\mathrm{l}}}{\rho_{\mathrm{l}}} \right) \right] , \tag{4.14}$$

where T_0 is the melting temperature of the metal, w_0 the specific heat per unit mass and R the particle radius. γ_{s} and γ_{l} are the specific surface energies and ρ_{s} and ρ_{l} the densities of the solid and liquid state, respectively.

For the upper bound of the melting temperature, the onset of thermodynamic instability has been used as the melting criterion. The maximum melting temperature $T_{\mathrm{R}}^{\mathrm{max}}$ is defined as

$$T_{\mathrm{R}}^{\mathrm{max}} = T_0 \left[1 - \frac{2\gamma_{\mathrm{sl}}}{w_0 \rho_{\mathrm{sl}} R} \right] , \tag{4.15}$$

where γ_{sl} is the surface energy between solid and liquid states, and ρ_{sl} the average of ρ_{s} and ρ_{l}. Since the specific surface energies and densities vary relatively little between the solid and liquid states, and since also $T_{\mathrm{R}} \propto R^{-1}$ in both equations, melting point depression is only relevant to smaller particle radii $R \leq 5$ nm. Other models describing melting temperature decrease in metal particles show qualitatively similar dependencies, as suggested in [409–413]. Values determined experimentally, for example, for Ag and Au particles [413] for Sn, Bi, In, and Pb particles [13], for Fe particles [409] or for Pb particles ($R \approx 10$ nm) in SiO_2 [410], lie below the bounds determined by (4.14) and (4.15).

Since the bulk melting temperatures for gold ($T_{\mathrm{M}} = 1\,337$ K) and silver ($T_{\mathrm{M}} = 1\,235$ K) [414] are relatively high, and the temperatures for thermal treatment are around 700 K, it is assumed that melting point depression is negligible during the thermal treatments of plasma polymer films with embedded silver or gold nanoparticles. Consequently, melting point depression cannot be responsible for nanostructural changes.

In the case of indium with its significantly lower melting temperature ($T_{\mathrm{M}} = 429.8$ K) [414], the minimum and maximum melting temperatures have been estimated for different mean particle radii using (4.14) and (4.15) and compared with experimentally determined values [13]:

$$R = 10 \text{ nm}, \quad 387 \leq T_{\mathrm{R}} \leq 416 \text{ K}, \quad T_{\mathrm{R}}^{\mathrm{exp}} = 415 \text{ K},$$
$$R = 5 \text{ nm}, \quad 340 \leq T_{\mathrm{R}} \leq 405 \text{ K}, \quad T_{\mathrm{R}}^{\mathrm{exp}} = 385 \text{ K}.$$

The experimental determination of melting point reductions for indium particles, produced by thermal evaporation and investigated in situ with the electron microscope, has been carried out by means of dark-field images during very slow thermal treatment ($1\,\mathrm{K\,min^{-1}}$). A melting temperature of 380 K was determined for indium particles ($R = 10$ nm) in an iron matrix [415]. Since the melting temperature of bulk indium is already in the thermal treatment temperature range for plasma polymer films with embedded indium nanoparticles, the melting point reduction of indium particles must be considered in any discussion of possible nanostructural changes, in contrast to the case for embedded gold or silver nanoparticles.

4.2 Thermal Treatment of Plasma Polymer Films with Embedded Nanoparticles

4.2.1 Change in Particle Size and Shape Distribution

Due to the relatively good thermal stability of the polymer matrix, as confirmed by optical investigations (see Sect. 2.1.3), particle-containing plasma polymer films were initially thermally treated for only a short time in high vacuum (10^{-3} Pa, 500–550 K), the aim being to realize changes in the particle size and shape distributions without destroying the multilayer system.

Figure 4.5 shows TEM micrographs for four samples of a plasma polymer multilayer system containing silver nanoparticles (film II/3) after film deposition. The metal content decreases continuously from Fig. 4.5a through g. The right-hand images of Fig. 4.5 are TEM micrographs for the four samples after a 15 min thermal treatment at 450 K. Since the samples located on the electron microscope grids were thermally treated ex situ in a high vacuum chamber after the first TEM micrographs were taken, the second set of TEM micrographs will not have been taken at exactly the same positions after the thermal treatment.

The changes in particle size and shape are shown in Fig. 4.6 with the appropriate three-dimensional particle size and shape distribution histograms. The mean particle sizes $\bar{D}$ and shape factors $\bar{S}$ are given in Table 4.2.

Table 4.2. Mean particle size and mean particle shape of embedded silver particles before and after thermal treatment. See also Figs. 4.5 and 4.6

	Before thermal treatment			After thermal treatment	
	$\bar{D}$ [nm]	$\bar{S}$		$\bar{D}$ [nm]	$\bar{S}$
(a)	46.0 ± 38.8	0.52 ± 0.22	(b)	41.6 ± 33.0	0.70 ± 0.13
(c)	25.6 ± 8.2	0.68 ± 0.17	(d)	31.6 ± 14.8	0.72 ± 0.13
(e)	20.1 ± 5.6	0.76 ± 0.12	(f)	24.6 ± 8.4	0.75 ± 0.13
(g)	16.1 ± 5.7	0.73 ± 0.19	(h)	18.8 ± 6.5	0.80 ± 0.10

Mean particle sizes changed after thermal treatment for samples (c), (e) and (f). The decrease in mean particle size for sample (a) results from the large number of small particles. Most of the embedded silver is distributed over a few very large particles. However, the smaller number of very large particles have almost no influence on particle size statistics, especially on the

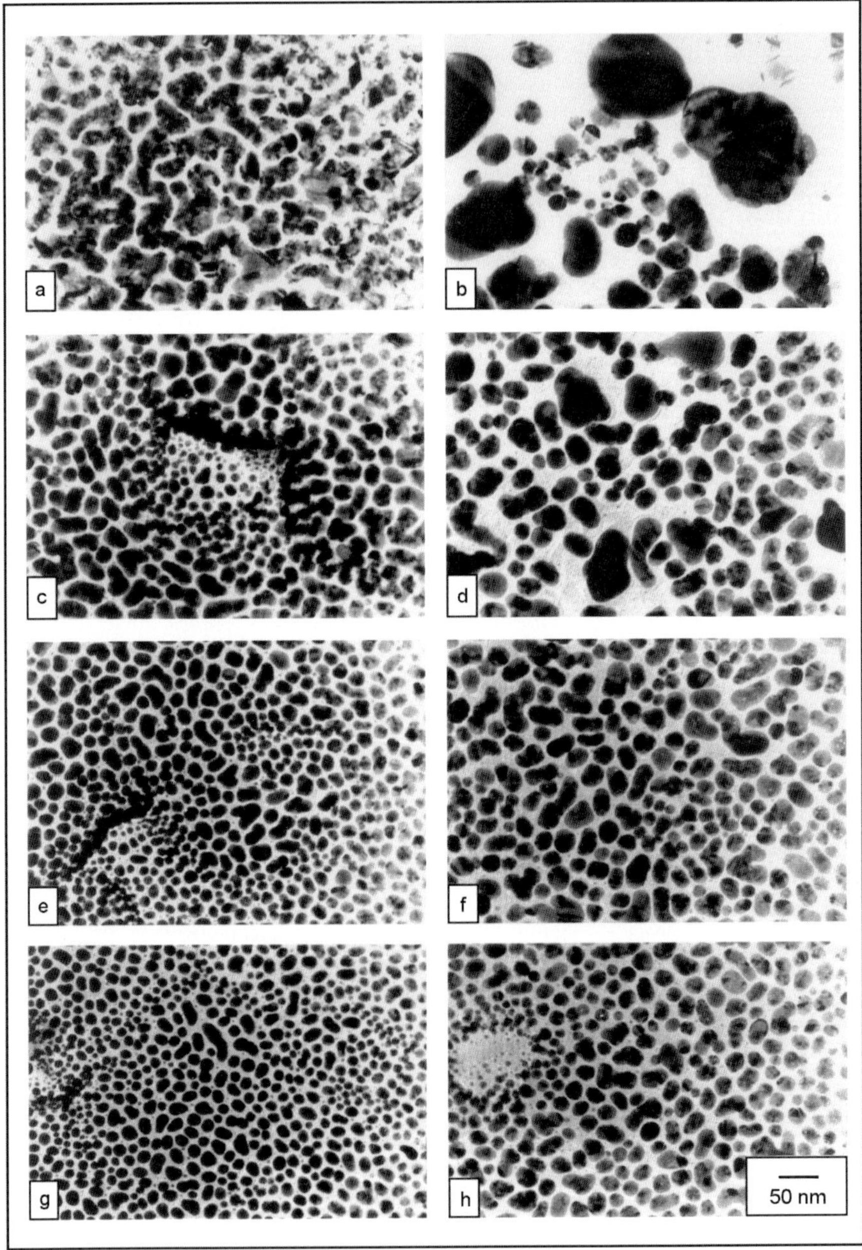

Fig. 4.5. (a–h) TEM micrographs for four samples of a plasma polymer multilayer with embedded silver nanoparticles (*left*) before and (*right*) after thermal treatment (450 K)

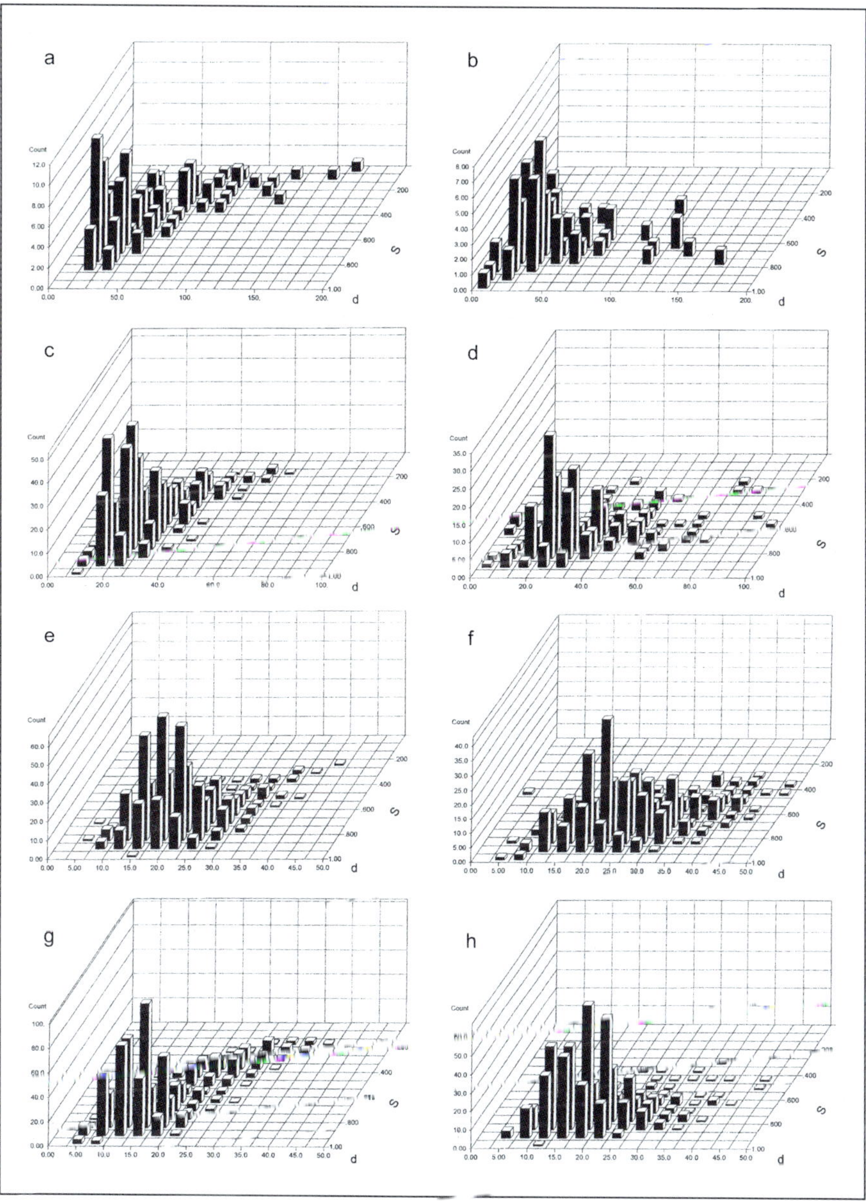

Fig. 4.6. (a–h) Three-dimensional particle size and shape distribution histograms for the TEM micrographs in Fig. 4.5

mean values for particle shape and size, compared with the effect of the many small particles. Very clear changes in the mean shape factor $\bar{S}$ occurred for sample (a). Before thermal treatment, silver particles with $\bar{S} = 0.52$ were observed, but the particles had a more spherical mean shape factor of $\bar{S} = 0.70$ after thermal treatment. The changes in particle size and shape are well demonstrated in the three-dimensional particle size and shape histograms for sample (a). The largest particles have a diameter of $D = 170$ nm and shape factors of $S \leq 0.1$ before thermal treatment. There is also a large number of particles with sizes in the range $80 \leq D \leq 130$ nm and shape factors $S \leq 0.5$. After thermal treatment, the few large particles have particle sizes in the range $90 \leq D \leq 160$ nm and shape factors $S \geq 0.5$ (with only one exception). Comparison of the histograms in Figs. 4.6a and b shows clearly that the thermal treatment is also responsible for the changes in the shape factor of the embedded silver particles which accompany the size changes. This effect is also observed for other samples. Comparison of the histograms in Figs. 4.6g and h shows a slight change in mean particle size from $\bar{D} = 16.1$ nm to $\bar{D} = 18.8$ nm, and also that particles with $S \leq 0.4$ no longer exist after thermal treatment.

The clear changes in particle size and shape seen in Fig. 4.5 are mainly caused by recrystallization and coalescence of the silver particles (Sect. 4.1.3). Furthermore, plasma polymer macromolecules located between two particles can be modified or destroyed by the thermal treatment, so that two particles that were previously separated by a thin intermediate layer can come into contact. It is not possible from the TEM micrographs of Fig. 4.5 to say whether that process actually took place. As can be seen from the particle size and shape distribution histograms of Fig. 4.6, the number of embedded particles decreases with reshaping and coalescence.

After thermal treatment the area filling factor decreases significantly, as can be seen from the values in Table 4.3.

Table 4.3. Area filling factor of the embedded silver particles before and after thermal treatment for the TEM micrographs in Fig. 4.5

Before thermal treatment	After thermal treatment
(a) $F_A = 0.89$	(b) $F_A = 0.32$
(c) $F_A = 0.91$	(d) $F_A = 0.74$
(e) $F_A = 0.76$	(f) $F_A = 0.51$
(g) $F_A = 0.88$	(h) $F_A = 0.61$

Since the multilayer system maintains its structure, as XPS investigation shows, there is no material loss from the silver particles. The decreasing area filling factor results from increased vertical particle size which is in turn caused by recrystallization and coalescence. This also leads to an increase in the composite layer thickness d_c. However, if the total film thickness of the multilayer system stays constant, then it can be assumed that changes take place in the film thicknesses d_{p1} and d_{p2} of the lower and upper plasma polymer layers. This change in the individual film thicknesses must be included, for example, in optical calculations (Sect. 6.2.4).

XPS depth profiles confirm that the multilayer system – first plasma polymer layer/plasma polymer layer with embedded silver nanoparticles/second plasma polymer layer – remains unchanged after thermal treatment. The left-hand diagram of Fig. 4.7 shows the XPS depth profiles of carbon, oxygen and silver for a multilayer system (film II/4) deposited with comparable parameters to the multilayer system in Fig. 4.5, apart from the total film thickness. The right-hand diagram of Fig. 4.7 depicts the depth profile for

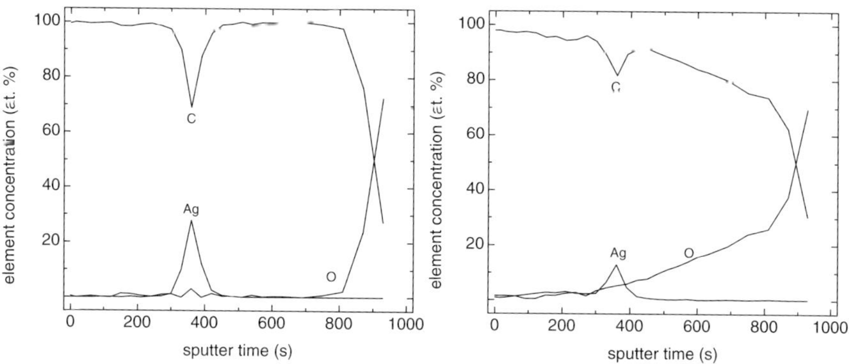

Fig. 4.7. XPS depth profile of a plasma polymer multilayer system with embedded silver nanoparticles (*left*) before and (*right*) after thermal treatment for 15 min up to 450 K

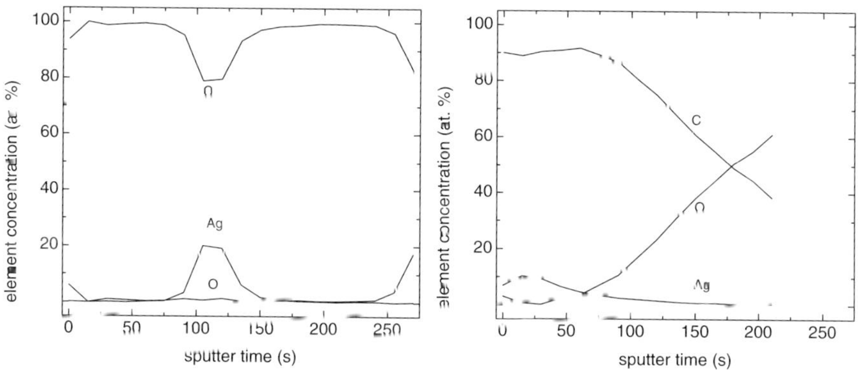

Fig. 4.8. XPS depth profile of a plasma polymer multilayer system with embedded nanoparticles (*left*) before and (*right*) after thermal treatment for 180 min up to 570 K

another sample of this film, treated under the same thermal conditions as the film of Fig. 4.5 (15 min, 450 K). The embedded silver particles are still located between the two plasma polymer layers, and the multilayer system is still intact. The increased oxygen proportion after sputter times of $t_{sp} = 400$ s can be explained by the micro-cracks which occur in the plasma polymer film after thermal treatment. Comparable observations of the formation of micro-cracks and fold-overs were obtained for plasma polymer films made from hexamethylcyclotrisilazane [124] after thermal treatment. Due to these film inhomogeneities, the substrate will be sputtered partly. Any oxygen diffusion from the SiO_2 substrate can be neglected. No shifts have been observed in the C $1s$ or Ag $3d$ binding energies.

Of course, it is possible to destroy the multilayer system with higher treatment temperatures and longer treatment times, as seen in Fig. 4.8. The left-hand diagram shows the depth profile of a plasma polymer multilayer system with embedded silver nanoparticles after fabrication (film II/4), whilst the right-hand image shows the situation after thermal treatment (570 K, 180 min). The shorter sputter times t_{sp}, compared to Fig. 4.7, result from an increased sputter energy, only used for that experiment. After thermal treatment a multilayer system was no longer detectable, and Ag $3d$ binding energies were detected on the film surface. It can be assumed that the plasma polymer matrix is destroyed and silver emerges partially at the surface. Furthermore, the carbon of the plasma polymer matrix is sputtered off much faster. Oxygen is detected after sputter times $t_{sp} \geq 90$ s, showing that there is no closed plasma polymer film on the wafer and SiO_2 layer. Analysis of the C $1s$ and Ag $3d$ binding energies did not indicate a chemical bond between silver and oxygen or carbon. These film samples, not located on copper grids for electron microscopy, were destroyed by the thermal treatment and could not be investigated with the TEM.

With ex situ thermal treatment, a change in particle size and shape distributions can be determined, and the stability of the multilayer system can also be investigated. However, information concerning diffusion processes or possible particle migration cannot be obtained. This is possible with in situ thermal treatment inside the electron microscope and presented in the next section.

In Situ Investigations During Thermal Treatment. Figure 4.9 shows TEM micrographs and three-dimensional distribution histograms for a plasma polymer multilayer system with embedded silver nanoparticles (film I/13) after thermal treatment in an electron microscope for 20 min (left) and 360 min (right). The temperature in the sample holder was determined as 775 K. However, it can be assumed that the real temperature in the investigated sample region is much lower. This is caused by the thermal stability and low heat conductivity of the thin plasma polymer film. The determination of temperature on free-standing thin films presents a general problem because of the complicated heat transfer processes in insulating materials with embedded metal

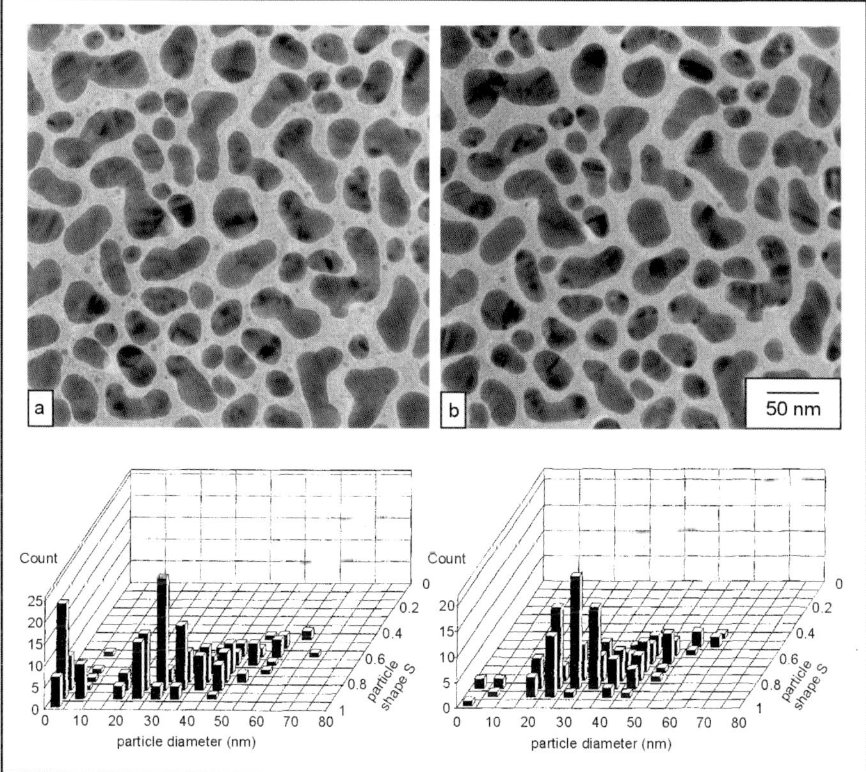

Fig. 4.9. TEM micrographs and three-dimensional particle size and shape distribution histograms for a plasma polymer multilayer system with embedded silver nanoparticles after thermal treatment of 20 min (**a**) and 360 min (**b**) at constant temperature in situ inside the electron microscope

particles. The thermal conductivity can be estimated using effective medium theories [416], although the approximation is very rough. This is why all nanostructural changes occuring in situ inside the electron microscope are discussed without knowledge of the true thermal treatment temperature at the sample in the electron microscope.

During thermal treatment, most small particles between the large silver particles disappear, as is also shown in the three-dimensional particle shape and size histograms. The number of particles with $D \leq 15$ nm drops from 65 to 6. The large particles have not changed their position. However, it can be assumed that the single crystallites of the polycrystalline particles have changed orientation, because contrast differences within particles look different.

Since coalescence or recrystallization of silver particles has obviously not taken place, the disappearance of the small silver particles can be explained by atomic diffusion through the polymer matrix (Ostwald ripening). The present investigations are not sufficient to prove the validity of the rate equation (4.4) or to determine the proportionality constant K_{LSW} in (4.5).

However, the particle size distribution can be correlated with the distribution function derived from LSW theory and with a statically determined distribution function for higher filling factors. Figure 4.10 shows the particle size distribution for the LSW theory, calculated using (4.6) ($f \approx 0$) and the particle size histograms from the TEM micrographs in Fig. 4.9. Neither histogram agrees well with the theoretical distribution function. This is a consequence of the many small particles evident before thermal treatment and the inadequate assumption $f \approx 0$. The particle distribution function for a two-dimensional diffusion (4.8) shows no better agreement with the experimental particle size histograms, because $f \approx 0$ is also applied for (4.8).

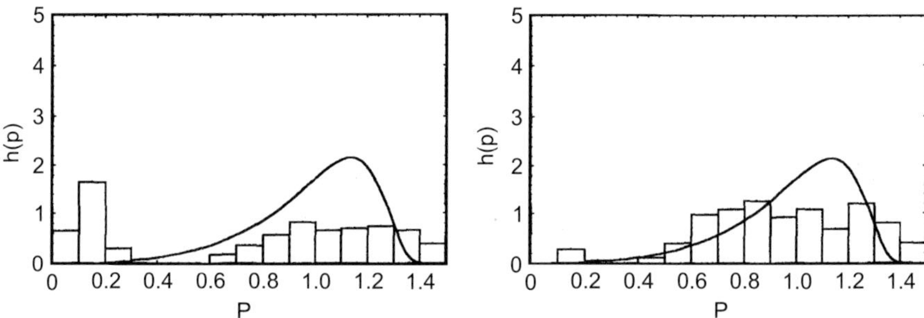

Fig. 4.10. Particle size distribution $h(p)$ according to LSW theory and particle size histograms for the TEM micrographs of Fig. 4.9

Figure 4.11 shows a correlation of the experimental particle size distribution after thermal treatment (Fig. 4.9b) with the particle size distributions for different filling factors after Voorhees and Glicksman [381] as depicted in Fig. 4.3. The agreement between the experimental histograms and the calculated particle size distribution is a little better. By means of the particle size distribution, a graphically determined filling factor of $f \approx 0.40$ is estimated from Fig. 4.11 for the sample of Fig. 4.9b. Since $f_{\mathrm{max}} = 2f_{\mathrm{A}}/3$ is applied for elliptical particles with uniform particle size, a maximum filling factor $f_{\mathrm{max}} = 0.41$ can be estimated from the area factor $f_{\mathrm{A}} = 0.61$ determined by image analysis of Fig. 4.9b, which corresponds to the graphically determined value from Fig. 4.11.

In situ thermal treatment inside the TEM also offers the possibility of studying the coalescence of two neighboring silver nanoparticles during the coalescence process. Much effort is required to ascertain temperature conditions under which coalescence takes place slowly enough for it to be observable it in several stages. The energy of the electron beam which can be focused and defocused also affects the coalescence velocity. It can be assumed that

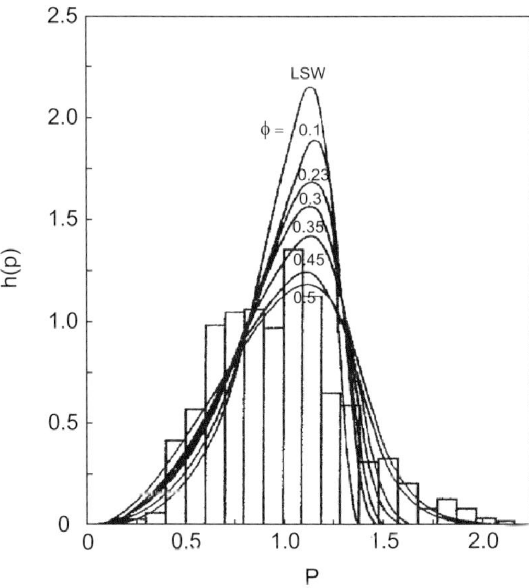

Fig. 4.11. Particle size distributions for different filling factors after [381] with the
particle size histogram from Fig. 4.9b

the focused electron beam acts as an additional heat source and that coalescence can be stimulated step by step. In Sect. 4.4.4, Fig. 4.28 illustrates the
coalescence of two neighboring silver particles under the influence of electron
beam irradiation.

Figure 4.12 shows the time progression of a nanostructural change during in situ thermal treatment inside the electron microscope, taking another
sample from the plasma polymer multilayer system with embedded silver
nanoparticles (film I/13). The temperature measured on the sample holder
was 275 K for Fig. 4.12a and 775 K, 875 K and 975 K for Fig. 4.12b, c and
d, respectively, but it must be assumed that the real sample temperature is
much lower. In contrast to the simultaneous electron irradiation described
in Sect. 4.4.3, the sample was only irradiated during thermal treatment if a
micrograph was to be recorded.

The few very small silver particles seen between the very large ones at the
beginning of the thermal treatment disappear as it proceeds. This indicates
atomic diffusion (Ostwald ripening) similar to that in Fig. 4.9. At the same
time, the polycrystalline structures of the silver particles change.

If thermal treatment of the sample is continued, there is a sudden disap
pearance of parts of or even whole silver particles (Fig. 4.9d). The contrast
change starts on the edge of the particle and proceeds over the entire particle
within a few seconds. The former shape of the particle is still visible by the
higher contrast in the polymer matrix on the former particle edge. It has been

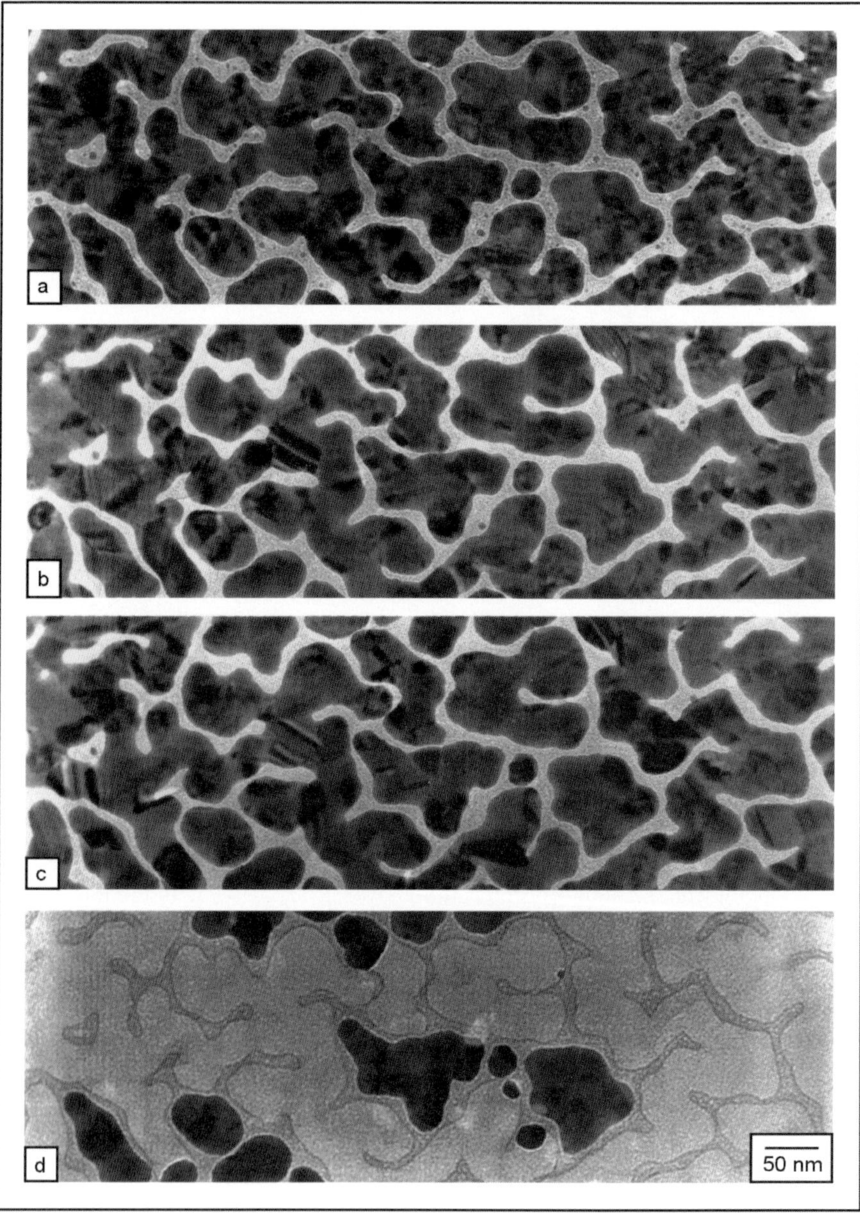

Fig. 4.12. (a–d) TEM micrographs of a plasma polymer multilayer system with embedded silver nanoparticles during in situ thermal treatment in the electron microscope

demonstrated by nanospot electron energy diffraction (EDX) that there is no silver in the bright regions of Fig. 4.9d.

There are two possibilities to explain the disappearance of silver particles. First, it can be assumed that micro-cracks develop in the plasma polymer matrix during thermal treatment in the electron microscope, and that the melted silver particle drops out from the self-supporting plasma polymer matrix. XPS investigations (Fig. 4.7) show some indications of these micro-cracks. The sublimation of the silver particle caused by additional thermal treatment from electron irradiation could be a more likely possibility. This sublimation of metal particles during electron irradiation, also investigated for gold particles in [417], assumes that the upper plasma polymer film of the multilayer system has disappeared by thermal desorption. At present no complete explanation is known for the disappearance of silver particles as shown in Fig. 4.9.

The outcome of the described TEM investigations is as follows. It can be assumed that changes in the size and shape distributions of embedded particles are mainly generated by reshaping and coalescence of the silver particles. Figures 4.9 and 4.12 indicate that atomic diffusion also occurs during thermal treatment. This can be described partly by the LSW theory of Ostwald ripening. Up to now, there is no evidence for metal particle migration during thermal treatment, but particle migration cannot be excluded generally.

It might be possible to bring about a desired nanostructural modification, if there were some way to limit the thermal treatment of the films to a very small substrate area. Since coalescence and reshaping of metal particles occurs in a very short time period and yields clear changes in the film nanostructure, the generation of spatially limited coalescence will be explained by means of a laterally resolved nanostructural modification in Sects. 4.3 and 4.4.

4.2.2 Particle Oxidation

For plasma polymer films with embedded indium particles, nanostructural changes comparable to the coalescence or recrystallization of silver particles were not observed. However, a completely different possibility was found for nanostructural change by thermal treatment. Figure 4.13 shows TEM micrographs for a sample of a plasma polymer multilayer system with embedded indium nanoparticles (film II/2) before (left) and after (right) ex situ thermal treatment at 10^{-2} Pa up to 500 K (20 min). This temperature is above the melting point of indium (430 K). The multilayer system was not destroyed, thanks to the good thermal stability of the polymer matrix. However, because of the low heat conductivity of the plasma polymer, it is not certain whether indium particles really anneal up to the melting point during thermal treatment. In fact, neither the mean particle size nor the mean shape factor change significantly during the thermal treatment. Furthermore, no significant changes in particle size and shape were observed from particle size and shape distribution histograms.

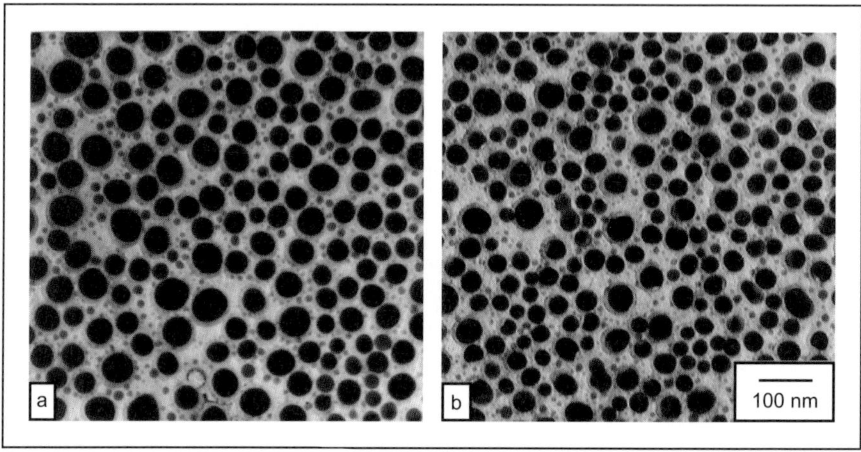

Fig. 4.13. TEM micrographs of a plasma polymer multilayer system with embedded indium nanoparticles (**a**) before and (**b**) after thermal treatment (20 min up to 500 K)

The nanostructural changes that took place in the film could only be shown by electron diffraction analysis. For better analysis and interpretation, the intensities of the diffraction rings were integrated and presented as a function of the inverse diameters. Maxima from the resulting radial diffraction intensity distributions can be mapped to the diffraction pattern of the embedded particles.

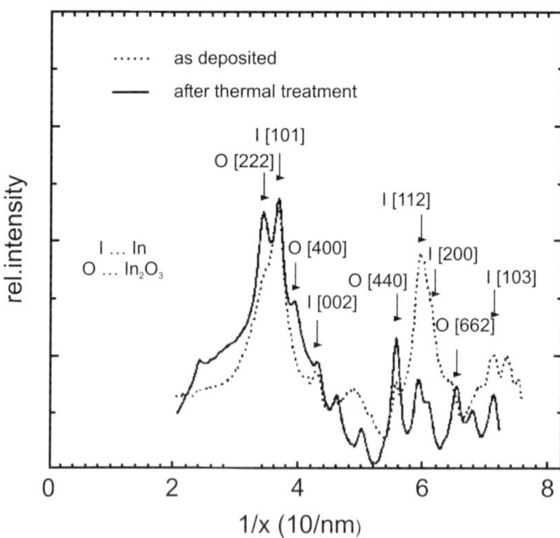

Fig. 4.14. Electron diffraction intensity over the inverse lattice plane spacing for the sample in Fig. 4.13

Figure 4.14 shows the radial diffraction intensity distributions for the sample in Fig. 4.13 before and after thermal treatment. In and In_2O_3 can be clearly distinguished owing to their different crystal structures (In is tetragonal and In_2O_3 is cubic). While almost no In_2O_3 reflections are observed before thermal treatment, their proportion is much higher after thermal treatment. It can therefore be assumed that In_2O_3 forms during thermal treatment. Since the particle size and shape distributions barely change, it can be concluded that In_2O_3 encapsulation occurs around the metal particle. This corresponds to the formation of oxide encapsulation explained in Sect. 4.1.5.

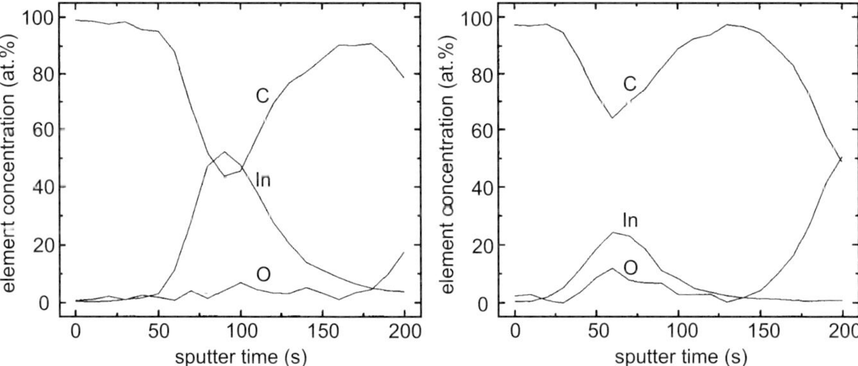

Fig. 4.15. XPS depth profile for a plasma polymer multilayer system with embedded indium nanoparticles. *Left*: sample before and *right*: sample after thermal treatment up to 500 K

XPS investigations should solve the question of where the oxygen comes from which is needed to oxidize the indium particles. Figure 4.15 shows the depth profiles of a plasma polymer multilayer system with embedded indium nanoparticles (film II/2) before and after thermal treatment (20 min, 500 K) under vacuum conditions ($\leq 10^{-2}$ Pa). It is confirmed that the multilayer system is maintained after thermal treatment. The slightly lower etching time to reach the substrate is mainly the result of the lower indium content of the thermally treated sample, but a lower thermal desorption of the plasma polymer matrix also contributes. In contrast to multilayer systems with embedded gold or silver nanoparticles, an oxygen proportion a clear 6–8% over the detection limit can be detected for plasma polymer multilayers with embedded indium nanoparticles before thermal treatment. The regions of the multilayer system without embedded indium contain no oxygen before or after thermal treatment. The increased oxygen proportion after sputtering times of $t_{sp} \geq 170$ s and $t_{sp} \geq 150$ s indicates arrival at the SiO_2 layer of the wafer. From this and from the peak shape of the C $1s$ binding energy at 285 eV, it can be concluded that the existing oxygen is bound to the indium. Moreover, investigation of chemical shifts in the In $3d_{5/2}$ binding energy at

444 eV and the In $3d_{3/2}$ binding energy at 452 eV has shown a significant proportion of In_2O_3, and this is increased after thermal treatment. The ratio of In_2O_3 to In increases from lower than 1:10 to about 1:3.

The indium oxide content may be developed during film deposition by the reaction between indium atoms and residual gas in the reactor, or by slow in-diffusion of oxygen at room temperature. Since the amount of oxygen contained in the film clearly increases during thermal treatment, there must be an increased in-diffusion of oxygen through the polymer matrix during or after thermal treatment. This was not observed for the gold and silver cases (see Fig. 4.7). The oxygen needed for oxidation during thermal treatment can come from the polymer surface, where oxygen is adsorbed, or from the residual gas during the vacuum thermal treatment at 10^{-2} Pa. Because of the high thermal stability of thermally oxidized SiO_2, the substrate can be ruled out as an oxygen source. The sample preparation for TEM investigations could be a further source of oxygen, since samples are dissolved in distilled water to detach the films from the substrate, and this may introduce OH groups. However, XPS investigations of samples deposited on thermally oxidized silicon wafers which did not come in contact with water have also shown an increased proportion of oxygen. This means that the brief water exposure of electron microscopy samples is only one of several oxygen sources.

The examples of thermally induced nanostructural changes in plasma polymer multilayers with embedded silver or indium nanoparticles show that it is possible to realize nanostructural changes without loss of mass or destruction of the multilayer system. This has advantages when investigating the dependence of optical properties on the particle size and shape distribution. Optical properties can be determined on one sample before and after thermal treatment, and correlations with the change in optical properties can be carried out without the complication of a mass loss. It is also possible to realize position-dependent nanostructural changes by thermal treatment, for example by laser or electron irradiation.

4.3 Laser Irradiation

Laser irradiation can cause changes within the film as well as a partial or complete ablation of the film itself. The laser irradiation of plasma polymer films with embedded nanoparticles can therefore be classified in terms of the two different types of nanostructural modification:

- laser irradiation with loss of mass of the plasma polymer,
- laser irradiation without loss of mass of the plasma polymer, but in which metal particle size and shape distributions are changed.

Figure 4.16 depicts the difference between these two possibilities. The plasma polymer is pyrolized by the local annealing of the laser irradiation with loss of material [267], and a thinner layer with an increased metal content remains.

In the ideal case, only a metal film remains. If a metal-containing film with a metal content below the percolation threshold is irradiated, then conductive paths can be generated [241, 242, 285]. This could lead to a lot of possible applications, for example, in the fabrication of microelectronic devices [240].

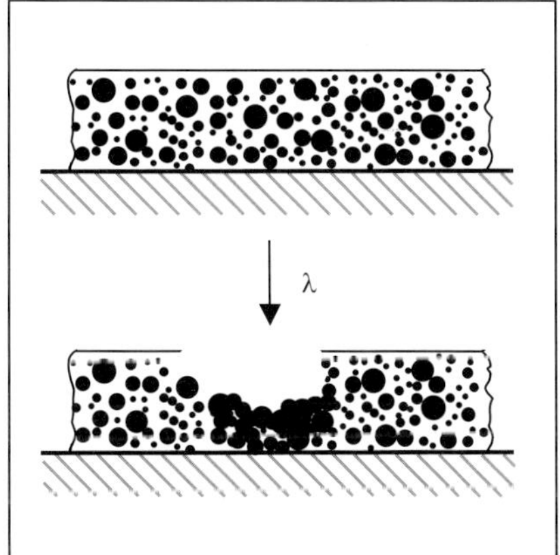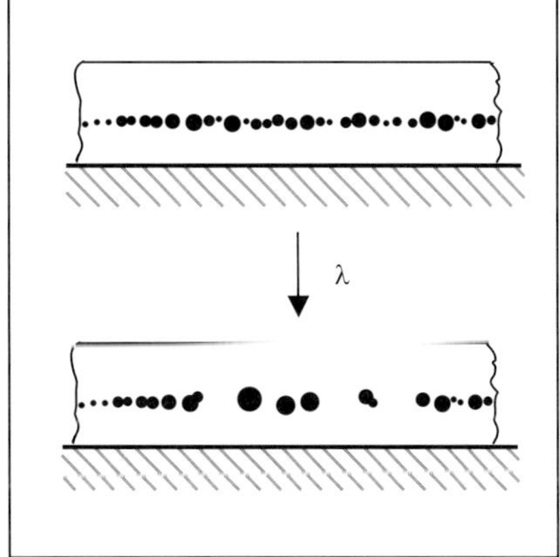

Fig. 4.16. Nanostructural changes in polymer films with embedded nanoparticles as a result of laser irradiation. *Left*: laser ablation of the polymer matrix and coalescence of the particles. *Right*: coalescence of embedded metal particles without modifying the polymer matrix

The second aim behind laser irradiation of plasma polymer films with embedded metal nanoparticles is to change the size and shape distributions of embedded metal particles in the laser irradiated region, without modifying the polymer matrix and without losing mass by ablation. Effective limitation of nanostructural changes to the irradiated substrate area requires the film to be heated directly through its optical absorption at the laser wavelength, and not through reflection or absorption by the substrate.

Plasma polymer films with different embedded metals deposited on quartz substrates were irradiated by a laser beam (Nd–YAG laser, 1064 nm, CW mode, laser power 8.6 W, beam diameter 100 μm) along a single line in order to determine the local expansion due to laser irradiation, and with a bar pattern (distance about 80 μm) over a substrate area of 10 × 30 mm for optical measurements. Modifications in the films generated by laser irradiation were visible to the naked eye. Metal-containing plasma polymer films on copper grids were also irradiated along a single line in order to determine local nanostructural changes for investigations with the electron microscope.

Figure 4.17 shows TEM micrographs for two samples of a plasma polymer multilayer system with embedded silver nanoparticles (film II/5) before laser irradiation (left) and after laser irradiation (right). Samples before laser irradiation show the nanostructures for metal proportions over the percolation threshold, with large incompletely separated silver particles (Fig. 4.17a). In

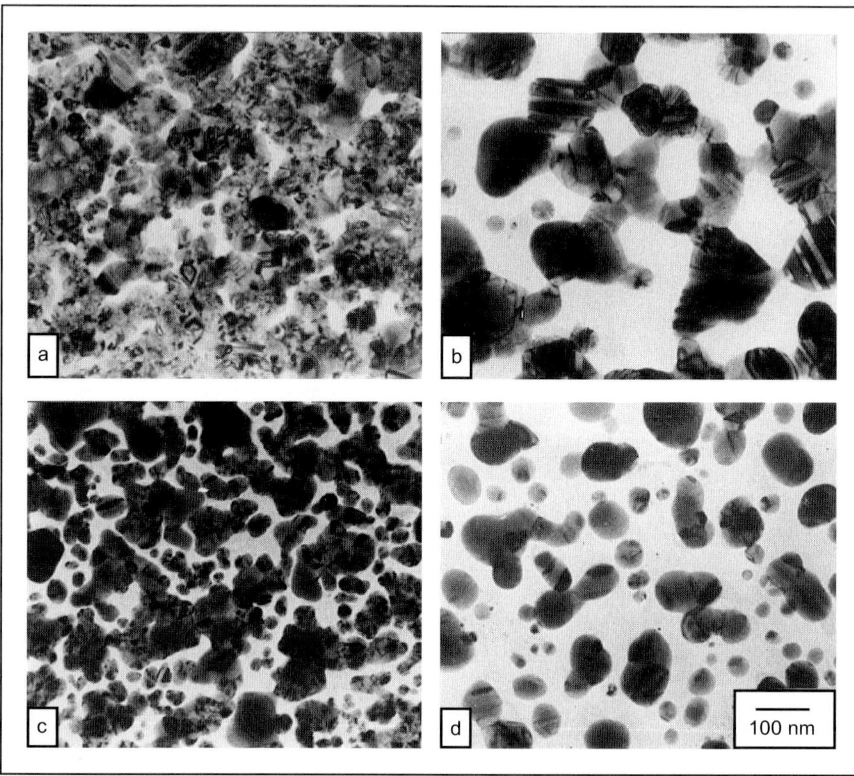

Fig. 4.17. (a–d) TEM micrographs of two samples of a plasma polymer multilayer with embedded silver nanoparticles (*left*) before and (*right*) after laser irradiation

contrast, after a little laser irradiation, the samples exhibit relatively large silver particles, in the range 100–200 nm (Fig. 4.17b). The filling factor for the area decreases from $F_A = 0.80$ to $F_A = 0.55$, but this decrease does not imply that a loss of mass has occurred. In fact, flat particles oriented parallel to the substrate develop a more spherical shape during laser irradiation, as with thermal treatment. This causes the apparent decrease in the silver area filling factor in lateral TEM micrographs.

A decrease in area filling factor is also observed for the second sample, with $F_A = 0.70$ going to $F_A = 0.45$. The TEM micrographs in Figs. 4.17c and d both clearly depict single particles, and analyses of particle size and shape distributions are possible. While the mean particle sizes $\bar{D} = 66 \pm 82$ nm and $\bar{D} = 64 \pm 52$ nm are not much different, due to the dominance of a few very large particles, the standard deviation is much lower after laser irradiation. Indeed, particle sizes $D > 200$ nm no longer occur, and very small particles with sizes $D < 5$ nm are seldom observed. The changes are more obvious in the shape factor S. While more than 20 particles had a shape factor of

$S \leq 0.3$ before laser irradiation, such low shape factors no longer appear after laser irradiation. Shape factor values show a peak at $S \geq 0.80$, and the mean shape factor increases from $\bar{S} = 0.59$ to $\bar{S} = 0.76$. The proportion of almost spherical silver particles has clearly increased.

Changes in particle size and shape distributions generated by laser irradiation correspond to the changes by thermal treatment as described in the previous chapter. It can therefore be assumed that local heating of the film by the laser beam leads to reshaping and coalescence of silver particles. This is also confirmed by traces of a thermal shock wave [330] visible in the electron microscope. Coalescence of metal particles without loss of mass induced by laser irradiation has also been observed, for example, in silver colloidal nanoparticles [418] or thin gold layers [399] and in Au–TeO$_2$ composite films [217]. Otherwise, the irradiation with femtosecond laser pulses of glass samples with embedded silver nanoparticles results in the creation of non-spherical particles from spherical particles with uniform orientation [419,420].

It has been confirmed by XPS investigations that there was no mass loss in the films during laser irradiation. The XPS depth profile of the Ag $3d_{5/2}$ and Ag $3d_{3/2}$ binding energies shows that the shift in the experimental position of the Ag binding energies results from an incompletely compensated surface charge whilst recording the depth profile. Figure 4.18 shows that the multilayer structure of the laser irradiated film (film II/5) remains intact. The curve near to the x–axis in the left-hand diagram is the XPS spectrum on the film surface at $t_{\mathrm{sp}} = 0$ s and the last line is the XPS spectrum at $t_{\mathrm{sp}} = 300$ s. In the right-hand depth profile, the curve near to the x–axis depicts the XPS spectrum after $t_{\mathrm{sp}} = 300$ s and the last line shows the XPS spectrum on the surface.

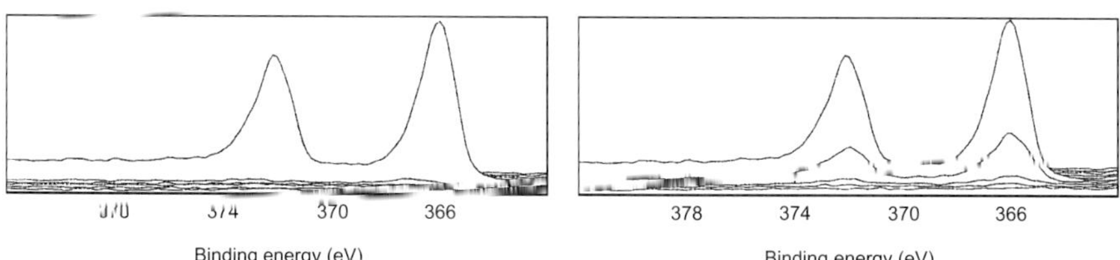

Fig. 4.18. XPS depth profiles showing Ag $3d$ binding energies from a laser irradiated sample of a plasma polymer multilayer system with embedded silver nanoparticles. *Left*: front view and *right*: back view of the silver particle layer inside the multilayer system

It can also be shown by UV visible–NIR spectroscopy that only changes in the particle size and shape distributions are induced by laser irradiation, and that the plasma polymer remains almost unchanged. As described in Sect. 2.1.3, optical transmission is strongly dependent on plasma polymer film thickness in the UV region. A decrease in film thickness induced by laser irradiation or a change in plasma polymer structure would be visible in the UV region of the optical transmission spectrum. Since no changes occur in this region (see Fig. 6.8 of Sect. 6.1.3), an almost unchanged plasma polymer film can be assumed.

Laser-induced local melting on the film surface provides a further possible form of modification. Similar investigations of surface modification by laser irradiation have been carried out on textured germanium films [421, 422] and textured tin films [423]. This melting on the film surface is also possible for plasma polymer films with embedded metal particles, since they also have a highly textured surface for large filling factors. This kind of surface modification has been realized by laser irradiation (ruby laser $\lambda = 694$ nm) on plasma polymer thin films with embedded germanium [325] or tin [316]. The aim in this work was to find material for holographic recording. Since only a low energy of $7 \times 10^{-3} \, \mathrm{J \, cm^{-2}}$ is required for a resolution of 2000 lines per millimeter, plasma polymers with embedded tin particles seem to be an appropriate material [316].

The results presented above have shown that local nanostructural changes are possible in plasma polymer films with embedded particles. Such changes can lead to coalescence or recrystallization of the embedded metal particles without changing the multilayer system. On the other hand, deliberate destruction of the polymer matrix or the textured film surface is also possible.

Spatially limited nanostructural changes in plasma polymer films with embedded particles have many possible applications, for example, in the fabrication of conductive metallic interconnects [240–242, 284] or materials for holographic recording [316, 325]. Because of the controllable change in surface roughness, its use as an intermediate layer to strengthen adhesion has been suggested in [330].

4.4 Electron Irradiation

4.4.1 Overview

Electron irradiation of plasma polymer films with embedded nanoparticles opens the way for lateral nanostructural modifications in the nanometer range by means of a change in the size and shape distributions of embedded metal particles, or their chemical structure. During experimental in situ analysis of the physical processes which take place, a method must be sought for achieving electron-beam-induced nanostructural changes which is compatible with microelectronics fabrication technologies. As for laser irradiation, a non-ablative way of modifying the nanostructure is required.

A nanostructural change can occur whilst films are being investigated in the electron microscope, by a simultaneous in situ thermal treatment. It is possible to carry out electron irradiation in the electron microscope, with simultaneous or subsequent investigation of irradiated samples. Investigation in the electron microscope requires a specific sample configuration, with samples placed on electron microscopy copper grids. Results must therefore be reinterpreted with respect to the electron irradiation delivered in an electron beam lithography system where the films would be deposited on silicon wafers, because of the different heat removal. From the standpoint of practical applications, however, irradiation in an electron microscope is only of academic interest.

In order to avoid this contradiction, an experimental system was set up for electron-beam-induced nanostructural change which guarantees electron irradiation from a small-focus electron source. Irradiated film samples can either be placed on electron microscopy copper grids or deposited on silicon wafers.

Nanostructural changes have been made by the following irradiation methods:

- electron irradiation in the electron microscope,
- electron irradiation in the electron microscope together with in situ thermal treatment,
- electron irradiation in an electron beam lithography system,
- electron irradiation with a small-focus electron source in a UHV system.

All four possibilities for electron irradiation offer irreversible nanostructural changes without mass loss.

The following laterally resolved irreversible nanostructural changes have been found as a result of electron irradiation of plasma polymer films with embedded nanoparticles (Fig. 4.19):

- electron-beam-induced metal oxidation – embedded indium particles are oxidized by electron irradiation in the electron microscope;
- electron-beam-induced out-hardening of the plasma polymer matrix – the plasma polymer matrix is hardened by electron irradiation in the electron microscope, whilst coalescence of embedded metal particles is impeded and takes place at higher temperatures;
- electron beam-induced metal particle coalescence – the coalescence of silver particles is initiated by electron irradiation from the small-focus electron source;
- electron-beam-induced metal particle coalescence and migration – besides metal particle coalescence, metal particle migration is caused by the electron irradiation, yielding a local change in the proportion of metal particles. An area with a metal particle deficiency is surrounded by areas of higher metal concentration.

These nanostructural changes and their irradiation methods are explained in detail in the following sections.

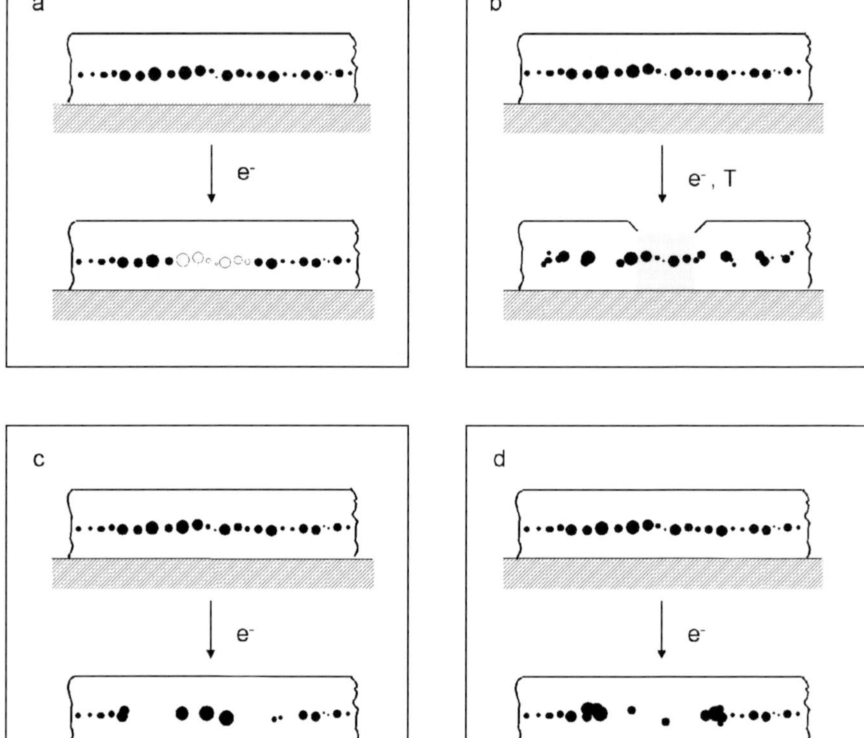

Fig. 4.19. Local modifications in the nanostructures of plasma polymer films with embedded nanoparticles, achieved by electron irradiation. (**A**) electron-beam-induced metal oxidation. (**B**) electron-beam-induced hardening of the plasma polymer matrix. (**C**) electron-beam-induced metal particle coalescence. (**D**) electron-beam-induced metal particle coalescence and migration

4.4.2 Electron Irradiation to Initiate Particle Oxidation

The previous discussion assumed that the interaction between electrons and sample in the electron microscope does not yield any changes in the nanostructure of nanoparticle-containing plasma polymer films. Generally, this is so for embedded silver and gold nanoparticles under usual irradiation conditions. Changes in the Bragg orientation of single crystallites of the polycrystalline metal particles which occur during investigations in the electron microscope result from very small changes in the position of metal particles within the polymer matrix. These can be generated by slight bending of the sample during irradiation and are not really nanostructural changes according to the definition of Chap. 1. However, nanostructural changes observed in plasma polymer films with embedded indium nanoparticles were directly induced by electron irradiation in the electron microscope.

During a normal electron microscopy investigation (TEM JEOL JEM 100 CX, 100 keV), the sample would be irradiated at a magnification factor of only 50 000 with about 50 nA μm^{-2} or about 500 pA μm^{-2} (TEM Philips CM20 FEG, 200 keV). Below these beam densities, changes in the sample are possible but very improbable, and cannot be observed. The current density of the beam can be increased by higher focusing. Beam current densities used for irradiation studies are about 2 μA μm^{-2} in the JEM 100 and about 500 nA μm^{-2} in the CM 20, whereby interaction with the sample is weaker for higher energy electrons. The nanostructural stability of plasma polymer films with embedded silver or gold particles deposited at power densities p $\geq$ 0.15 W cm^{-2} has been confirmed for these electron beam intensities. Nanostructural changes in the shape of embedded silver or gold particles, or coalescence of such particles generated only by electron irradiation in the electron microscope, were not detected, as has been observed for gold particles supported on SiO$_x$ under 3 keV electron irradiation [124].

Nanostructural changes in plasma polymer multilayers with embedded indium nanoparticles were obtained at higher focus using the JEM 100 electron microscope. However, satisfactory image production was not possible at higher focus because of beam inhomogeneities. Consequently, samples were irradiated instead at low focus with a higher beam current density ($\approx$ 2 μA μm^{-2}) and with a larger beam diameter at lower beam current density ($\approx$ 50 nA μm^{-2}). Time-resolved changes in the sample during irradiation with a focused electron beam were recorded by appropriate choice of irradiation time with a higher beam current density, and the progression of electron-beam-induced changes in the nanostructure were thereby monitored.

Figure 4.20 shows the change in the nanostructure of a plasma polymer multilayer system with embedded indium nanoparticles (film I/4) generated by increased beam current density. Figure 4.20a depicts the sample before the first irradiation with increased beam current intensity. Indium particles were observed with particle size and shape distributions comparable to those in Fig. 3.10c. The TEM micrographs of Figs. 4.20b and 4.20c were recorded after 6 min and 15 min irradiation, respectively, with higher beam focus. After 6 min, a strong decrease in contrast was already observed for some particles, continuing throughout irradiation. The contrast between particles became stronger than inside the particles. The previous indium particles were replaced by small, elongated particles forming at the former indium particle/polymer matrix interface.

Figure 4.21 shows a sample of the same film (film I/4) as in Fig. 4.20 after 15 min electron irradiation at the higher beam intensity. There are indium particles in the film with unchanged particle shape, as well as small crystallites formed by irradiation. Because of the higher resolution of Fig. 4.20 compared to Fig. 4.21, it can be seen that there are many small crystallites

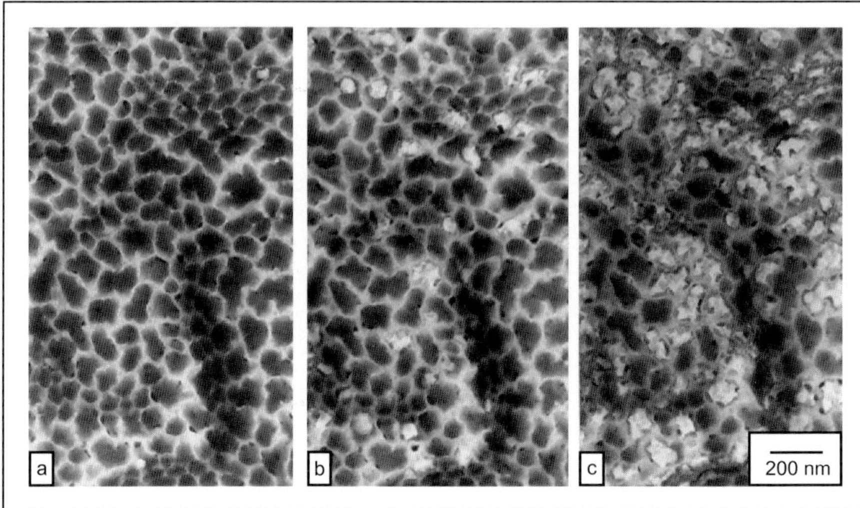

Fig. 4.20. TEM micrographs of a plasma polymer thin film with embedded indium nanoparticles (**a**) before, (**b**) after 6 min, and (**c**) after 15 min of electron beam irradiation with increased beam focusing

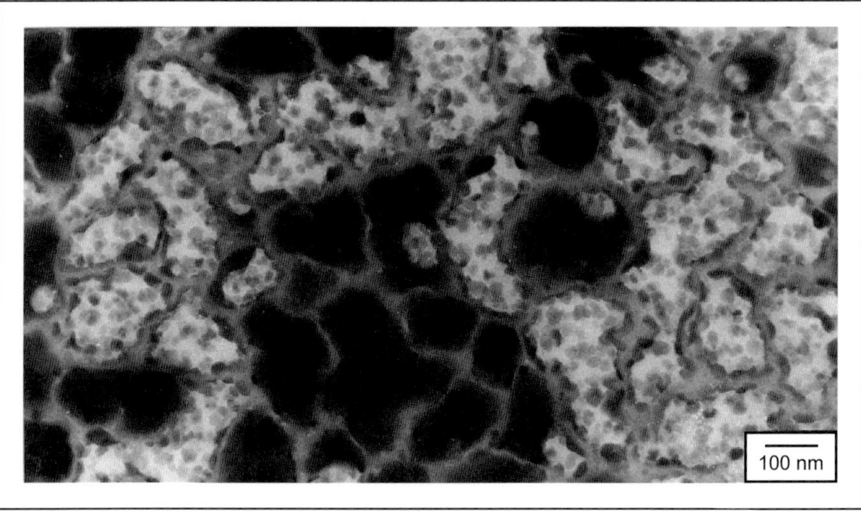

Fig. 4.21. TEM micrograph of a plasma polymer film with embedded indium nanoparticles after 15 min electron irradiation with increased beam focus

in the area previously occupied by a single indium particle. For a few indium particles, the nanostructural changes are still incomplete. While the indium particles possess sizes of $100 \leq D \leq 200$ nm before irradiation, the crystallites generated by irradiation are significantly smaller, with $10 \leq D \leq 20$ nm.

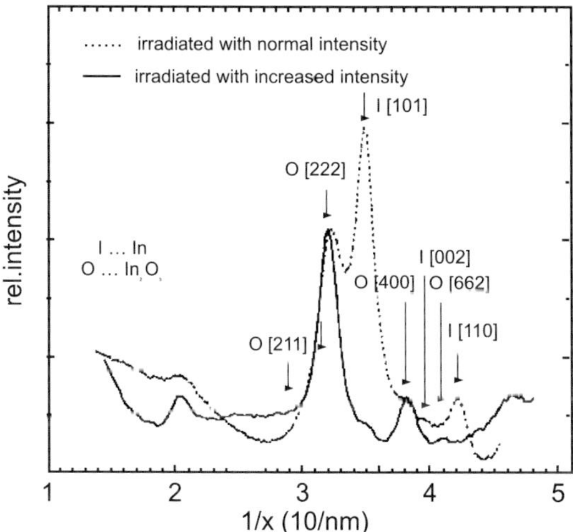

Fig. 4.22. Intensity over the inverse lattice plane distance for a sample irradiated with normal and with increased beam current density for 30 min (sample from Fig. 4.20)

Analysis of the electron diffraction pattern indicates that this process of nanostructural change occurs through oxidation of indium particles. Figure 4.22 shows the radial diffraction intensities of the film from Fig. 4.20 for the sample area which was irradiated with increased (30 min) and normal electron beam current densities. In the non-irradiated area, there are mainly In reflections, with little In_2O_3 content, while there are almost only In_2O_3 reflections within the irradiated sample area. Since the In and In_2O_3 reflections can be clearly distinguished because of their different lattice types, it can be assumed with confidence that the small crystallites are In_2O_3 particles, generated by electron irradiation with increased beam current density.

The extent of oxidation is also demonstrated in Fig. 4.23 for a sample of a plasma polymer multilayer system with embedded indium nanoparticles (film II/2). This shows the TEM micrographs (JEM 100) after an irradiation of (a) 3 min, (b) 6 min, (c) 9 min, (d) 12 min and (e) 15 min at higher beam current density. The progression of indium oxide formation could thus be observed. The two indium particles marked by arrows in Fig. 4.23a have a strong diffraction contrast before irradiation. Oxidation sets in after 6 min for the right-hand indium particle, spreading from the particle edge. The contrast change in the particles almost always starts at the edge of the particle,

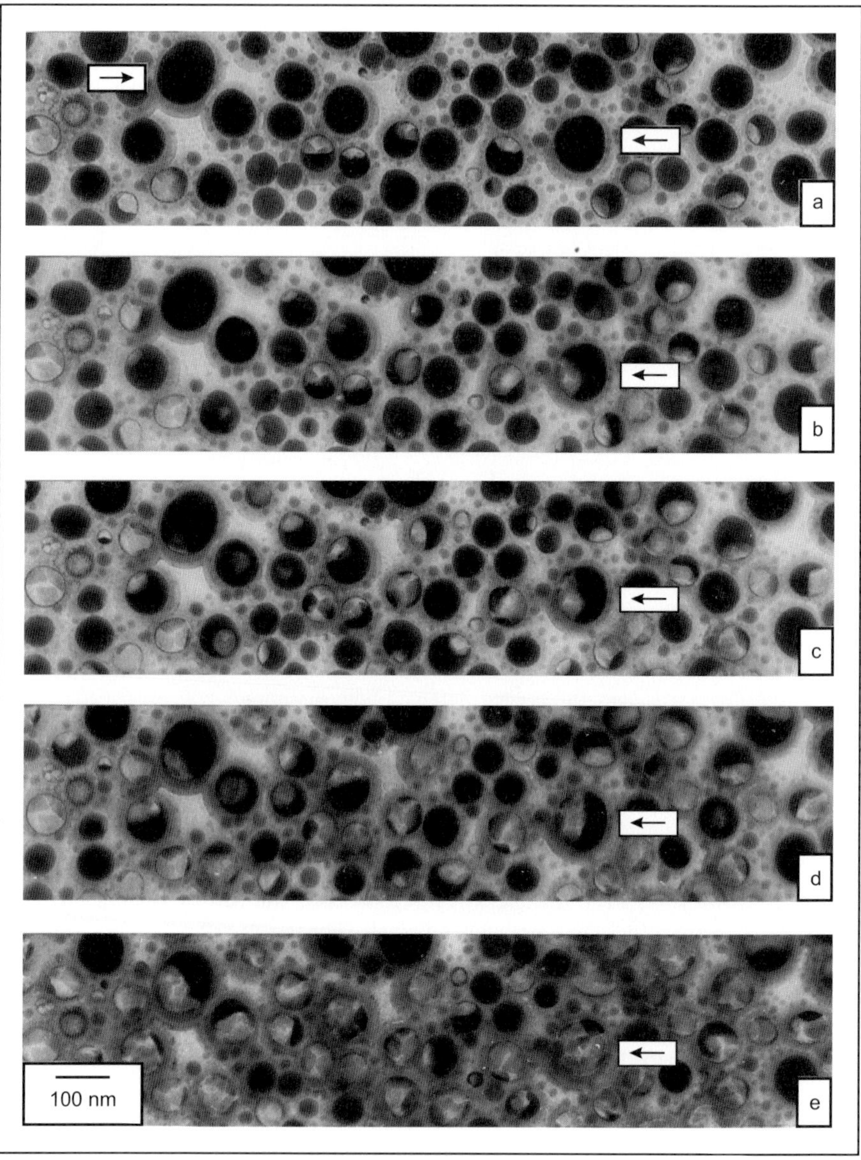

Fig. 4.23. (a–e) TEM micrographs of the oxidation sequence for embedded indium particles

as can also be seen for the left-hand particle in Fig. 4.23c. Starting from the particle edge, oxidation proceeds slowly through the entire (formerly indium) particle. Many small In_2O_3 particles (6–9 nm, 10–20 nm) are spawned from the large ($D = 80$ nm) indium particle (right arrow). A residue of the indium particle still exists, and the contours of the former indium particle are still recognizable. During irradiation with increased beam current density, no movement of the particles was observed in the polymer matrix. Furthermore, many small particles remain grouped around the large indium particles and it is not certain whether they consist of indium or indium oxide.

The upper part of Fig. 4.24 shows an indium particle from the sample of Fig. 4.23 at high resolution before electron irradiation with increased beam current density. The almost circular particle is surrounded by other particles at the edge of the picture. Between these particles are very small (D $\approx$ 2 nm) crystalline areas, also represented at atomic resolution, which can be attributed to the very small particles surrounding the larger particle. The lower section of Fig. 4.24 shows the crystal structure of an indium oxide particle formed after electron irradiation with increased beam focusing. Oblong particles developed. However, their lattice planes could not be assigned clearly to an indium oxide crystal structure.

XPS investigations of Fig. 4.15 showed that the plasma polymer multilayer system with embedded indium nanoparticles already contains a small proportion of oxygen after film fabrication. Contamination of the sample with oxygen from the residual gas in the electron microscope must also be considered as a further source of oxygen, apart from the possibilities already discussed in Sect. 4.2.2. As a reference point, the particle oxidation mentioned was only observed in the JEM 100 electron microscope, operated with an oil diffusion pump at residual gas pressures of 10^{-4} Pa. No particle oxidation was found in the CM 20 electron microscope, operated at ultra high vacuum with a residual pressure of 10^{-7} Pa and an ion getter pump. Point heating of the film by electron irradiation can also lead to intensified oxygen diffusion.

In contrast to the oxidation of indium particles by thermal treatment, which led to the formation of an oxide coating, electron irradiation with increased beam current density produces substantially smaller indium oxide particles. This completely different mechanism for the formation of indium oxide, which has so far only been found for plasma polymer films with embedded indium nanoparticles, makes it possible to generate indium oxide in a region limited to the illuminated sample area of the polymer matrix. Since this oxidation process proceeds slowly, the possibility exists of an in situ investigation of the oxidation process for nanoparticles

4.4.3 Electron Irradiation and Simultaneous Thermal Treatment to Initiate Coalescence and Reshaping

For the changes described so far in the nanostructure of plasma polymer films with embedded metal nanoparticles, no modification of the plasma polymer

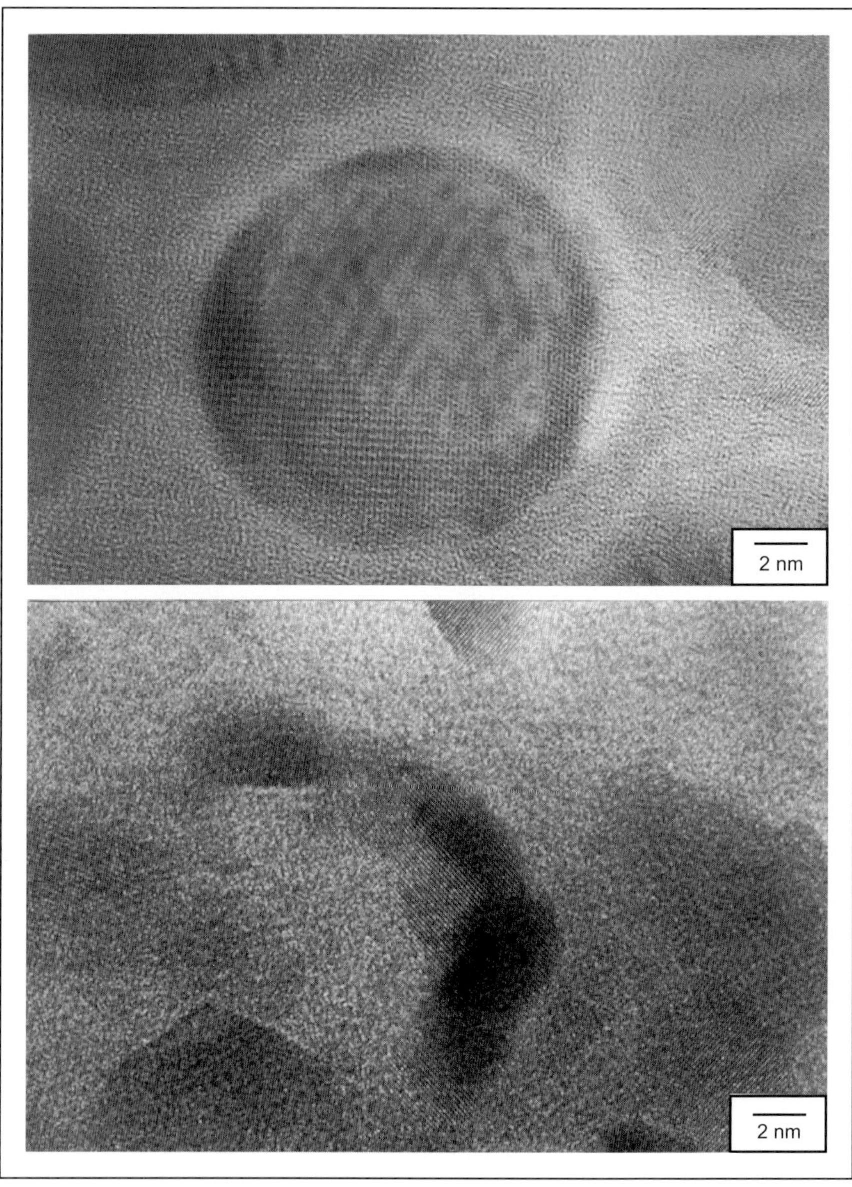

Fig. 4.24. TEM micrographs of an indium particle at high resolution before (*top*) and after (*bottom*) electron irradiation with increased beam focusing

matrix has been considered. In studies for the application of plasma polymer films as an electron beam resist, modifications were generated in the plasma polymer layer which then led to a selective behavior for the following etching procedure.

Electron irradiation of polymers leads to the formation or cracking of chains, or to reactions which close chains. These processes always take place at the same time, although one dominates. During the electron irradiation of plasma polymer films, further polymerization of the plasma polymer due to the available radicals and breakage of polymer chains are both possible. While modifications caused by electron irradiation are readily quantifiable in bulk materials, specific modifications in plasma polymer films can be determined only roughly, since relatively little information on the film structure can be obtained from IR spectroscopy and XPS investigations (see Sect. 2.1.3). Thus for example, information obtained on irradiation-induced polymerization and decomposition of styrene polymers is not generally transferable to plasma polymer layers.

Furthermore, hydrocarbons in the residual gas may be cracked by the electron beam effect, and a contamination layer containing carbon can form. This contamination layer leads to additional scattering of electrons and to a contrast reduction in the electron micrograph. It is specifically observed after a few seconds growth during investigation of films in the JEM 100 electron microscope, which is operated with a diffusion pump at a residual gas pressure of 10^{-4} Pa. For investigations in the CM 20 electron microscope, operated with an ion getter pump, the formation of a contamination layer is almost insignificant. Moreover, sample contamination is negligible at temperatures above 450 K.

Figure 4.25 shows a sample of a plasma polymer multilayer system with embedded silver nanoparticles (film I/13) after in situ thermal treatment in the CM 20 electron microscope. The thermal treatment (15 min, temperature on the sample holder 770 K) was carried out simultaneously with irradiation of the sample (beam diameter 20 µm, beam current density 500 pA µm^{-2}).

After thermal treatment, a darker circular area is found at the place on the sample where the electron beam was positioned during thermal treatment. The top part of Fig. 4.25 shows the edge of this irradiated area (irradiated area on the left and non-irradiated area on the right). Enlargements of sample areas of the irradiated and non-irradiated films are represented in the lower part of the figure. Nanostructural modifications of the silver particles within the non-irradiated sample area result from thermal treatment, mainly by coalescence and recrystallization

The silver particles have the same size and shape distributions as before thermal treatment in the irradiated sample area. Simultaneous electron irradiation during thermal treatment has prevented nanostructural modifications. This can be explained by a hardening of the plasma polymer matrix.

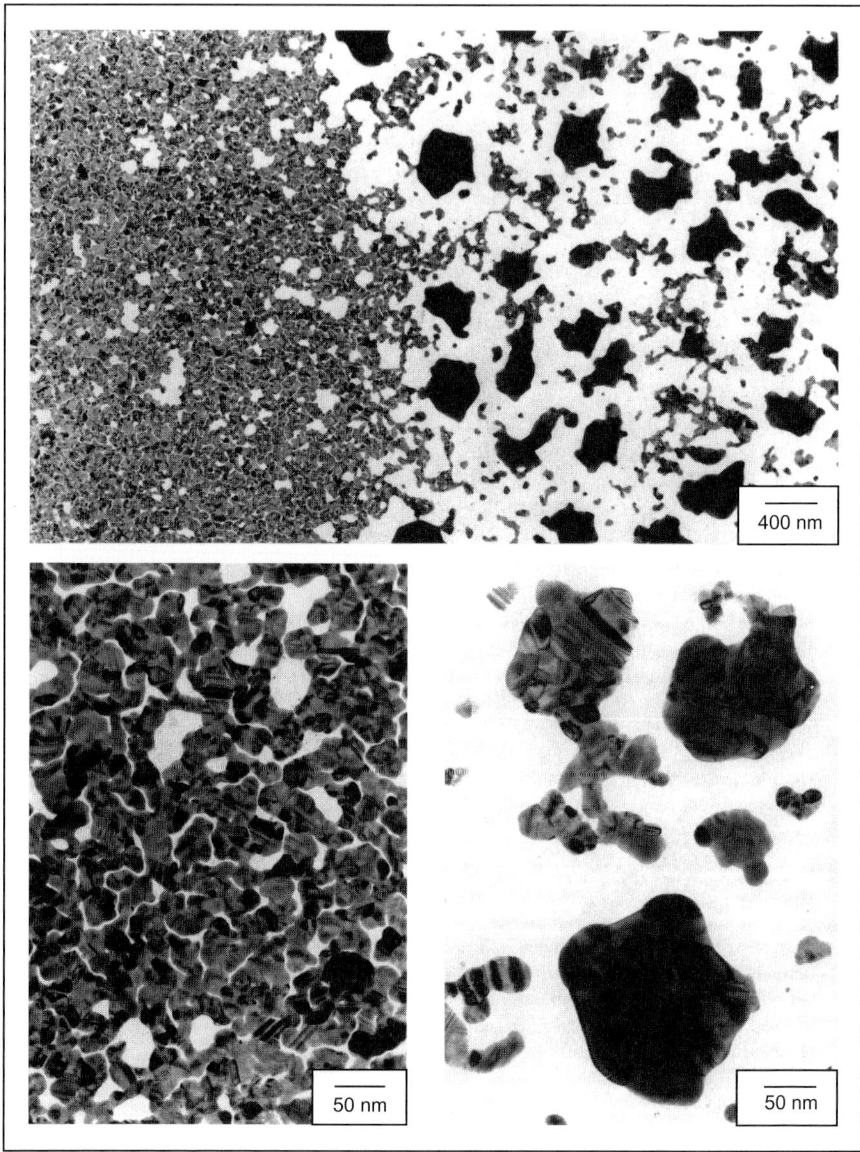

Fig. 4.25. TEM micrograph of a plasma polymer multilayer with embedded silver nanoparticles after thermal treatment. The *left-hand region* was electron-irradiated during thermal treatment. *Lower pictures*: section enlargements

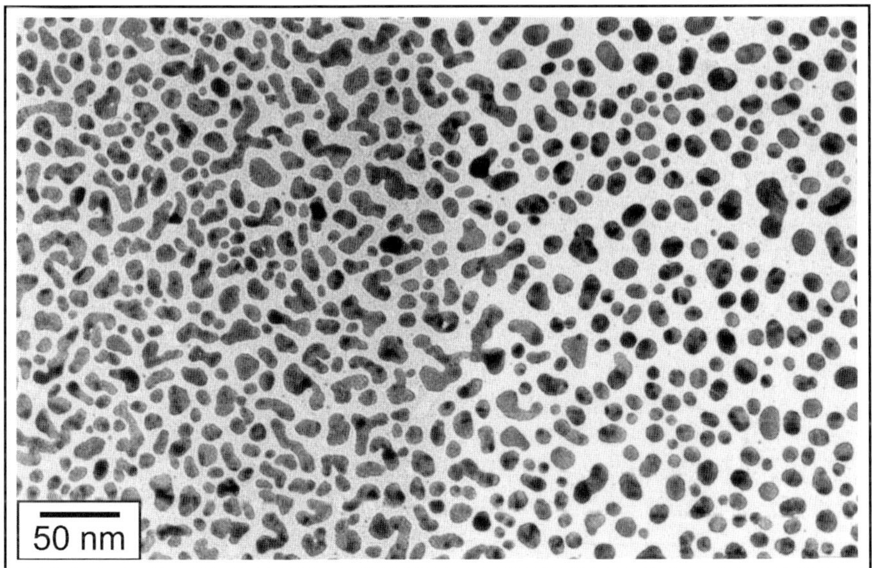

Fig. 4.26. TEM micrograph of a plasma polymer multilayer with embedded silver nanoparticles after thermal treatment. The *left-hand part* of the picture was electron-irradiated during thermal treatment

Figure 4.26 is a TEM micrograph of a sample that was thermally treated in the electron microscope. The electron beam was aimed at the left-hand part of the sample during the in situ thermal treatment. The maximum temperature on the sample holder was 770 K, but due to the low thermal conductivity of the sample and its low thermal capacity, it must be supposed that the real temperature in the investigated sample region was much lower.

Contamination and further hardening of the polymer matrix due to simultaneous electron beam irradiation and thermal treatment caused the nanostructure of the irradiated sample part to remain as deposited, even after thermal treatment. The right-hand part of Fig. 4.26 shows the nanostructure of the sample region without electron beam irradiation during thermal treatment. In this sample region, coalescence and recrystallization of silver particles took place. The left- and right-hand regions of Fig. 4.26 thus show silver particle sizes and shapes before and after nanostructural changes, respectively.

In order to get quantitative results concerning the size and shape distribution of embedded particles before and after thermal treatment and to obtain data for optical calculations with the Rayleigh–Gans theory (Sect. 6.2.2), particles are assumed to be elongated spheroids (rotationally symmetric ellipsoids). These are described by the semi-major axis A and semi-minor axis B. In total, 368 particles were analyzed, both before and after nanostructural changes. Results are given in three-dimensional histograms (Fig. 4.27).

The left-hand part of Fig. 4.27 shows the histogram for the sample as deposited and the right-hand part the histogram for sample regions on which nanostructural changes took place. The x direction gives the minor axis B, the y direction the major axis A, and the z direction the number of particles found in the A–B interval (count). The histogram for the sample as deposited (Fig. 4.27 left) shows a particularly large number of big particles, with eccentricity $e = B/A \ll 1$. It is clear that thermally induced coalescence and recrystallization have caused particle eccentricities to increase and particles to approach a spherical shape ($A \approx B$).

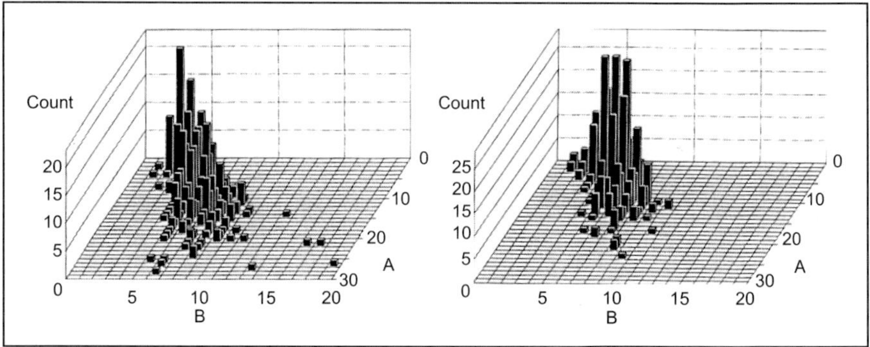

Fig. 4.27. Histograms for the semi-major and minor axes of the silver particles in Fig. 4.26. *Left*: for particles as deposited. *Right*: for particles after coalescence and reshaping

Figures 4.25 and 4.26 demonstrate that stabilization of the plasma polymer matrix is possible by subsequent hardening and that solidification of the plasma polymer matrix makes it necessary to increase the temperature to achieve any modification in the size and shape distributions of silver particles. This opens the way to a new method for nanostructural modifications in metal-containing plasma polymer films, combining electron irradiation and thermal treatment [239].

4.4.4 Electron-Beam-Initiated Coalescence

Up to now, the coalescence of two neighboring silver nanoparticles has not been observed by thermal treatment inside the transmission electron microscope because of the difficulty in establishing temperature conditions under which coalescence will take place slowly enough to be able to observe it dynamically. In fact, the energy from the electron beam, which can be focused and defocused, affects the coalescence velocity. It will be demonstrated that the focused electron beam acts as an additional heat source and that coalescence can be stimulated in a step-by-step manner. Due to the heat input from the electron beam, the plasma polymer matrix is stable enough to prevent destruction of the multilayer system but not strong enough to hinder coalescence.

Figure 4.28 shows the coalescence of two neighboring silver particles embedded in a plasma polymer multilayer (film I/12). The sequence starts with the as-deposited structure in which silver particles are well separated. After a first irradiation with the focused electron beam, a neck forms between neighboring particles in the middle of the picture. The neck increases and reaches the diameter of the smaller particle by the fourth picture. During coalescence, atomic planes are visible in the particles and also in the contact region. After defocusing the electron beam, coalescence stops.

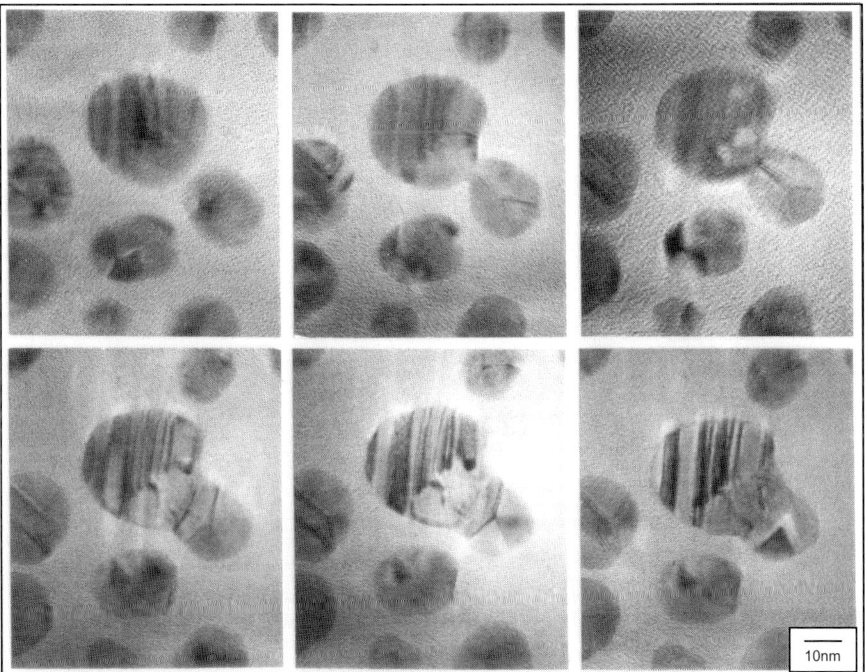

Fig. 4.28. Sequence of TEM micrographs during coalescence of two neighboring silver particles from *top left* to *bottom right*

All the images in Fig. 4.28 were taken at atomic resolution. During the experiments, atomic planes were visible at all times. The crystal structure remained and no melting of silver particles occurred. Atomic diffusion along the particle surface and along grain boundaries are entirely responsible for nanostructural changes. If the atomic planes are observed in more detail, it can be seen that their orientation changes during coalescence. Figure 4.29 shows a section enlargement of the region between the coalescing particles.

Atomic planes are marked with broken lines. During coalescence, the large particle on the left changed its crystal orientation. If the angle between the left border of the picture and the plane is measured, it changes from 65° in Fig. 4.29a to 114° in Fig. 4.29f. In addition, grain boundaries inside the particle change their orientation. Molecular dynamics computer simulation were employed to investigate the sintering mechanism of copper nanoparticles in [425]. It was calculated that misaligned particles rotate until they found a low-energy boundary. For two copper nanoparticles, a rotation angle of about 17° relative to each other was determined.

Figure 4.29 displays the first TEM investigation in atomic resolution dealing with the coalescence of embedded metallic nanoparticles. In [426], different steps of coalescence of gold particles supported on carbon films were observed but atomic resolution was not reached. The coalescence of silver nanoparticles on Ag(111) was also studied using AFM, but their crystal structure could not be observed [427]. Investigations with a transmission electron microscope having a DC-sputtering attachment demonstrated a reorientation on Ag particles on Cu(001) during thermal treatment [428].

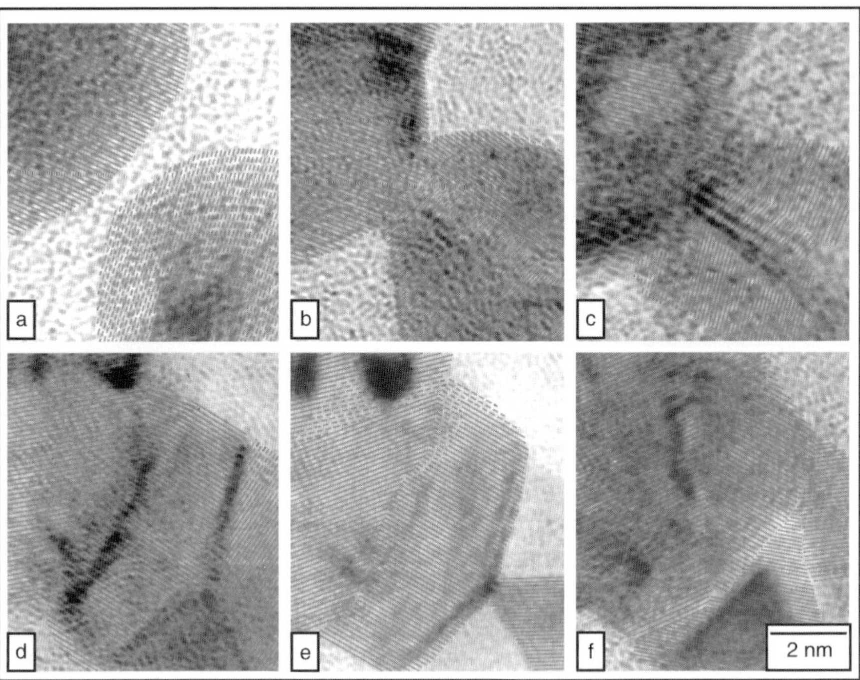

Fig. 4.29. Section enlargement of Fig. 4.28. Lattice planes are indicated by *broken lines*

4.4.5 Electron Beam Lithography

Since electron irradiation in the TEM can only be applied to thin films on copper grids, further methods must be developed for irradiating films on other substrate materials. One possibility has resulted from the use of a commercial electron beam lithography system (Jenoptik ZBA 21). Metal-containing films were located on $4''$ silicon wafers. Then, plasma polymer thin films with embedded silver or copper nanoparticles were irradiated with a 20 keV electron beam. The beam was widened to a maximum area of 6.3×6.3 μm^2. The beam current density was 3 A cm^{-2} at an irradiation dose of 2×10^{-3} C cm^{-2}. Sample areas of up to 3×3 mm^2 were irradiated step by step by moving the wafer. After irradiation, no changes were observed in the films. Irradiated sample areas became visible in the optical light microscope after thermal treatment (15 min, 550 K) [429].

Irradiation in the electron microscope cannot really be compared with irradiation in the electron beam lithography system. This is because heat is removed in a different way due to the different substrates (self-supporting film on copper grids and adhesive film on silicon wafer), and because there is electron back-scattering from the substrate material for films on silicon wafers. Since electron microscopy studies of nanostructural modifications produced by electron beam lithography and thermal treatment are not very practicable, this line of investigation was not pursued any further. However, it was possible to demonstrate in principle the feasibility of metal-nanoparticle-containing plasma polymer films as electron beam resist materials via irradiation in the electron beam lithography system.

The nanostructural modifications carried out so far have taken place in either the analysis system (a transmission electron microscope) or a commercial electron beam lithography system. In the TEM, basically only films on the usual electron microscope grids can be modified, and other types of substrate are not applicable for film irradiation. On the other hand, only normal silicon wafers could be irradiated in the commercial electron beam lithography system, and nanostructural analysis is not possible for these films since they cannot be removed from the silicon wafer.

It was therefore necessary to implement an experimental setup with a small-focus electron gun in order to permit the electron irradiation of films on any substrates. Irradiation with the small- or fine-focus electron gun forms a bridge between electron microscopy and electron beam lithography. Electron irradiation experiments on films with a small-focus electron gun had two aims: to implement lateral nanostructural modifications and to change the size and shape of single particles.

Figure 4.30 shows the UHV system designed for electron irradiation of plasma polymer films with embedded nanoparticles. A small-focus electron source was mounted horizontally inside a UHV chamber. To operate the small-focus electron gun, an operating pressure below 10^{-5} Pa must be ensured. This operating pressure is achieved with a turbo-molecular pump. A

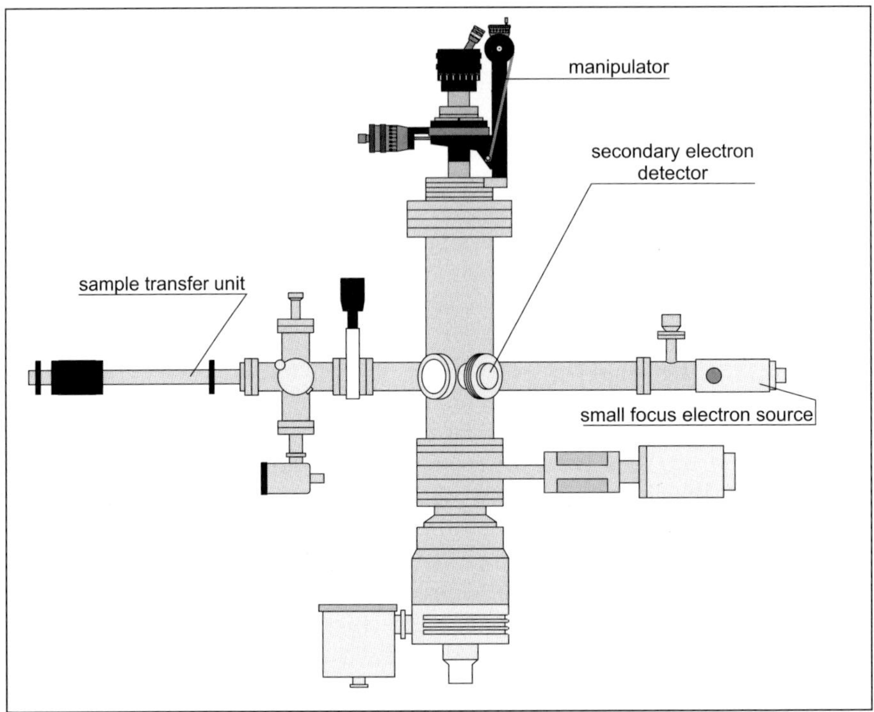

Fig. 4.30. UHV equipment for irradiating plasma polymer films with embedded nanoparticles using a small-focus electron source

secondary electron detector was built in at an angle of 60° to the electron gun to detect secondary electrons emitted over a two-dimensional sample region, with the assistance of a scanning generator. This signal is converted into a TV signal and presented on a monitor. A fluorescent screen was also used temporarily. The electron energy of the small-focus electron gun can be adjusted over the range 6–12 keV, and the diameter of the electron beam is ≥ 200 nm. The minimum electron beam diameter is achieved at a distance of 10–15 nm from the emission aperture. Rough mechanical adjustments and control of the primary power supply are made via the sample current measured at the sample holder. Typical values for the sample current during electron irradiation are 30 nA with a beam diameter of 5 µm. Fine focusing was done with the condenser.

The sample is brought into the work position of the electron gun (y direction) with a high precision manipulator. The sample is positioned in the x direction with a constant velocity stepping motor, guaranteeing irradiation in a bar pattern. Several of these line-like irradiations of the sample can be made by shifting the sample in the vertical z direction. An airlock with sample transfer mechanism enables fast replacement of samples. A special sample holder permits irradiation of up to 19 of the copper grids. Samples on silicon wafers can also be illuminated by the small-focus electron gun under the same conditions as for other sample holders.

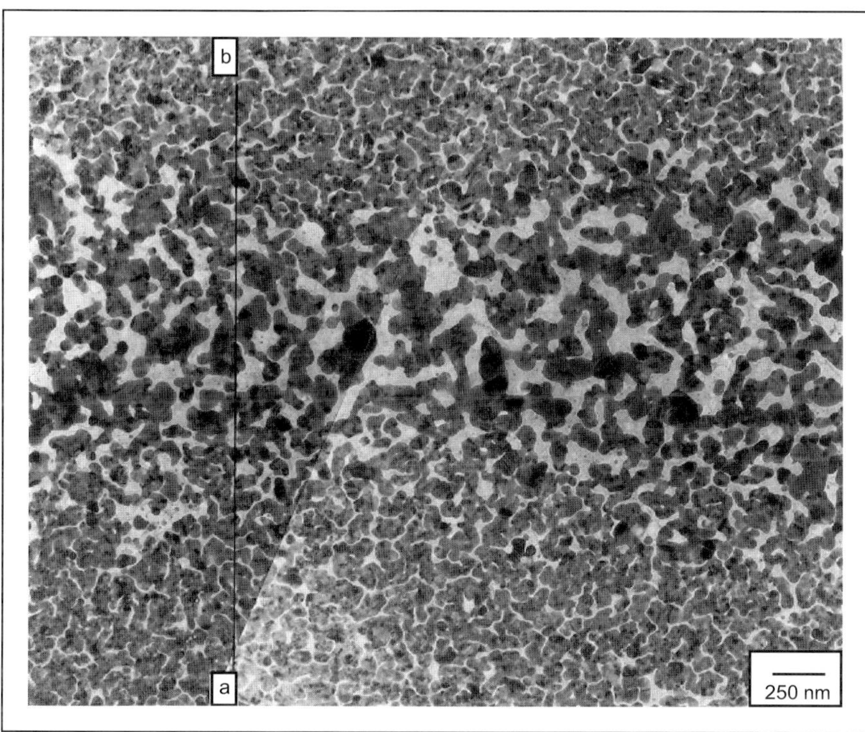

Fig. 4.31. Electron-beam-induced changes in the nanostructure of silver nanoparticles embedded in a plasma polymer thin film with embedded silver nanoparticles

Figure 4.31 shows the nanostructural changes of a plasma polymer multilayer system with embedded silver nanoparticles (film I/14) generated by irradiation using the small-focus electron source. The irradiation took place with a beam current of ≈ 50 nA and a beam diameter of ≈ 2 μm, and the sample was moved with a velocity of 100 μm s^{-1}. The electron-irradiated sample area is located in the center of the picture. Non-irradiated sample areas in the upper and lower parts show the particle size and shape distribution following film fabrication. Due to the electron irradiation, larger silver particles, which are still connected together, form by coalescence and reshaping. There are some very small ($D < 20$ nm) silver particles in the gaps between large silver particles. The electron-beam-modified sample area has a width of ≈ 2 μm, corresponding to the beam diameter, demonstrating that heat removal in the film is low, and that laterally limited nanostructuring is possible.

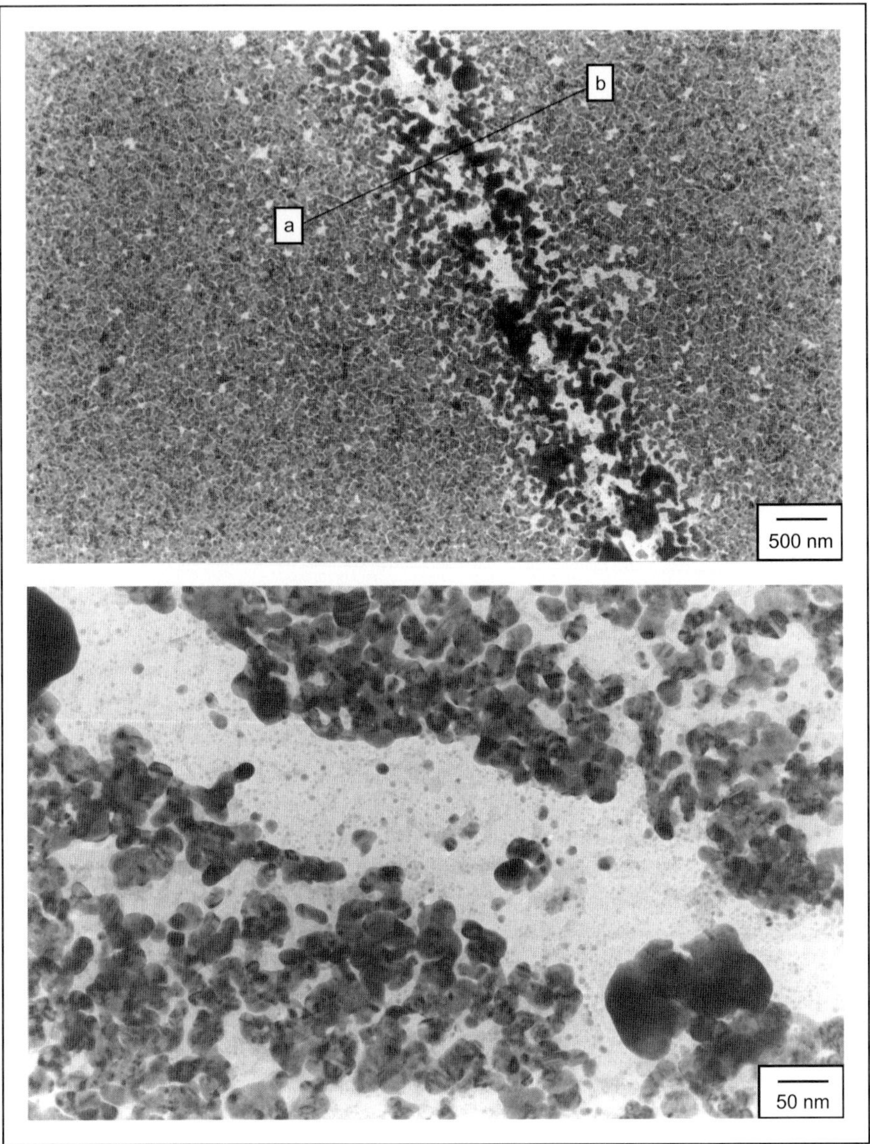

Fig. 4.32. Electron-beam-induced nanostructural changes in silver nanoparticles embedded in a plasma polymer multilayer. *Top*: overview. *Bottom*: section enlargement

Figure 4.32 shows a further stage of electron-beam-induced coalescence of the silver particles on a sample also taken from film I/14 (Fig. 4.31). The lower section of Fig. 4.32 shows an enlargement of the electron-beam-modified sample area, viewed from above. This area now has a slightly larger width of 2–3 μm. The beam diameter remained unchanged at about ≈ 2 μm, but the velocity of the electron beam scan was reduced to 50 $\mu m\,s^{-1}$ to double the energy input. The aggregated silver particles are now substantially larger and also spherical. There is an area of the sample of width approximately 300 nm in which no large silver particles have formed. The enlargement shows that a large number of very small silver particles ($D \leq 10$ nm) remain in this sample area. From the presence of these many small particles it can be assumed that the polymer matrix still exists and that the depletion area is not due to rupture of the plasma polymer film. Such a rupture has been found with further increased energy input.

By means of Figs. 4.31 and 4.32, two possible nanostructural modifications by electron-beam-induced coalescence can be distinguished. To do so, metal proportion modifications are considered along the sectional line from point (a) to point (b), across the width of the irradiated area perpendicular to the irradiation direction. With lower energy input, coalescence of metal particles is initiated. However, there was no modification in the metal proportion along the section from (a) to (b) in Fig. 4.31. In Fig. 4.32, at higher energy, two maxima of the metal content are observed at the edges of the irradiated area, with a clear minimum between them, as one traverses from (a) to (b). Consequently, particle migration took place in addition to silver particle coalescence for electron irradiation as seen in Fig. 4.32. Silver particle migration takes place towards the non-irradiated sample area. It is not clear, however, whether the particles move as such in the plasma polymer m atrix. An oriented coalescence away from the heated sample area is also possible. A distinction would only be possible by in situ investigations in the electron microscope. This, however, is hardly feasible because of the rapid progression of coalescence. The energy input needed to start coalescence or silver particle migration depends on both the size of the embedded silver particles and the stability of the surrounding plasma polymer matrix.

Electron irradiation with a small-focus electron source leads to immediate coalescence and reshaping of embedded silver particles. Depending on the energy input, coalescence and reshaping with and without particle migration can be distinguished. If simultaneous particle migration takes place, sample areas can be achieved which are almost silver-free.

The experimental UHV system used for electron irradiation with a small-focus electron gun provides for new experiments involving the irradiation of films on any substrate material. For example, it is possible to investigate changes in the electrical conductivity of metal-containing plasma polymer films during electron irradiation, or to modify the refractive index in a defined sample area and investigate optical wave guides in thin films. Furthermore, the experimental system developed also makes it possible to write regular structures into plasma polymer films with embedded metal particles without material loss, that is, by merely modifying the size and shape distributions of embedded metal particles. This means that new applications are possible as a result of the specific nanostructure and the modification of metal-containing plasma polymer films.

4.5 Nanostructural Changes Without Thermal Treatment

In order to complete the discussion, brief mention should be made of the possible influence of aging processes on nanostructural modifications in plasma polymer films with embedded metal nanoparticles.

Plasma polymer multilayers with embedded metal nanoparticles generally show very good long-term stability. No signs of aging were observed in plasma polymer layers made from fluorocarbons with embedded gold particles [256]. Plasma polymer films with embedded silver nanoparticles also show very good long-term stability if the metal particles are completely embedded in the plasma polymer matrix in the multilayer system. Optical measurements on these films over several years did not expose any significant modifications in transmission spectra [430]. Only plasma polymer layers made at lower power densities ($p \leq 0.1 \mathrm{~W\,cm}^{-2}$) have shown signs of aging in the plasma polymer, and also changes in the size and shape distributions of embedded metal particles by light exposure after some months due to the high proportion of free radicals. Silver sulfide formation was observed six months after film deposition for films with silver particles incompletely embedded in the polymer matrix [176]. Detachments due to inner stresses in the polymer matrix were not observed for the usually smaller film thicknesses $d \leq 500$ nm of multilayer systems.

It should be mentioned that similar structures to those seen in Fig. 4.32 can also be observed for the coalescence of silver particles during film deposition [306]. There are sample areas here with very small particles ($D \leq 20$ nm), where spherical silver particles ($D \approx 500$ nm) form at the edges. These aggregates of large silver particles, which are even observable in the optical light microscope, develop during film deposition at metal proportions close to the percolation threshold, probably by local discharges on the substrate surface.

It is further assumed that coalescence at room temperature with very long coalescence times (see Sect. 4.1.3) is without any importance for a defined nanostructural modification, because there are generally only a few days or weeks between film fabrication, nanostructural changes and electron microscope investigation.

Plasma polymer films with higher tin content have shown a whisker growth [336]. The films have a highly structured surface similar to plasma polymers with high indium content (see Fig. 3.22). Tin whiskers grow out from the film surface through the plasma polymer surface layer of the multilayer system after only few weeks. The whiskers have lengths up to several millimeters at diameters of a micrometer and are visible to the naked eye.

None of the room temperature changes in the nanostructure described briefly above influence our earlier conclusions concerning thermally induced nanostructural modifications.

5. Electronic Properties

5.1 Electronic Properties of Insulator Films with Embedded Metal Nanoparticles

Conductivity in the Three Structural Regions. The electrical direct current conductivity of a polymer film with embedded nanoparticles varies between the conductivity of the insulator material (e.g., for plasma polymers $\sigma_{po} = 10^{-14}$–$10^{-16}\ \Omega^{-1}\,\mathrm{cm}^{-1}$, see Sect. 2.1.3) and the bulk metal (e.g., for silver $\sigma_{Ag} = 6.3 \times 10^{5}\ \Omega^{-1}\,\mathrm{cm}^{-1}$, and for gold $\sigma_{Au} = 4.8 \times 10^{5}\ \Omega^{-1}\,\mathrm{cm}^{-1}$ [414]). Because this variation ranges over 20 orders of magnitude and no unified description of electrical transport processes exists for all possible metal proportions, different models are assumed for the d.c. electrical conductivity in each of the three structural regions: the metallic range, the percolation range and the insulating range (Fig. 5.1).

A continuous metallic layer with insulator inclusions exists in the metallic structure range ($f > f_{c}$). Particularly for high filling factors ($f \gg f_{c}$), films can be considered to be thin, continuous metallic films, and corresponding conductivity models can be applied. These models are based on the decrease in mean free path of conduction electrons due to dimensional constraints (Fuchs–Sondheimer theory). Above the Debye temperature, the linear dependence of the electrical resistance on temperature leads to a positive temperature coefficient of resistance (TCR) $\alpha = \delta R / R \delta T > 0$ [431].

The transition from a nearly metallic conducting layer to an insulating one takes place in the percolation range ($f \approx f_{c}$). The d.c. electrical conductivity is influenced by minimal changes in the metal proportion.

For a description of d.c. electrical conductivity in the dielectric structure range, a model of thermally activated tunneling between two neighboring, but separated metal particles is often invoked. The TCR is negative ($\alpha \leq 0$). The electrostatic activation energy can be calculated for approximately spherical particles from the particle size and spacing, and can also be determined experimentally by measuring the temperature dependence of the d.c. conductivity.

For very small particles, the electronic band splitting which results in quantum size effects has to be considered. The influence of quantum size effects on the electrical properties of particle assemblies is discussed in [9, 10],

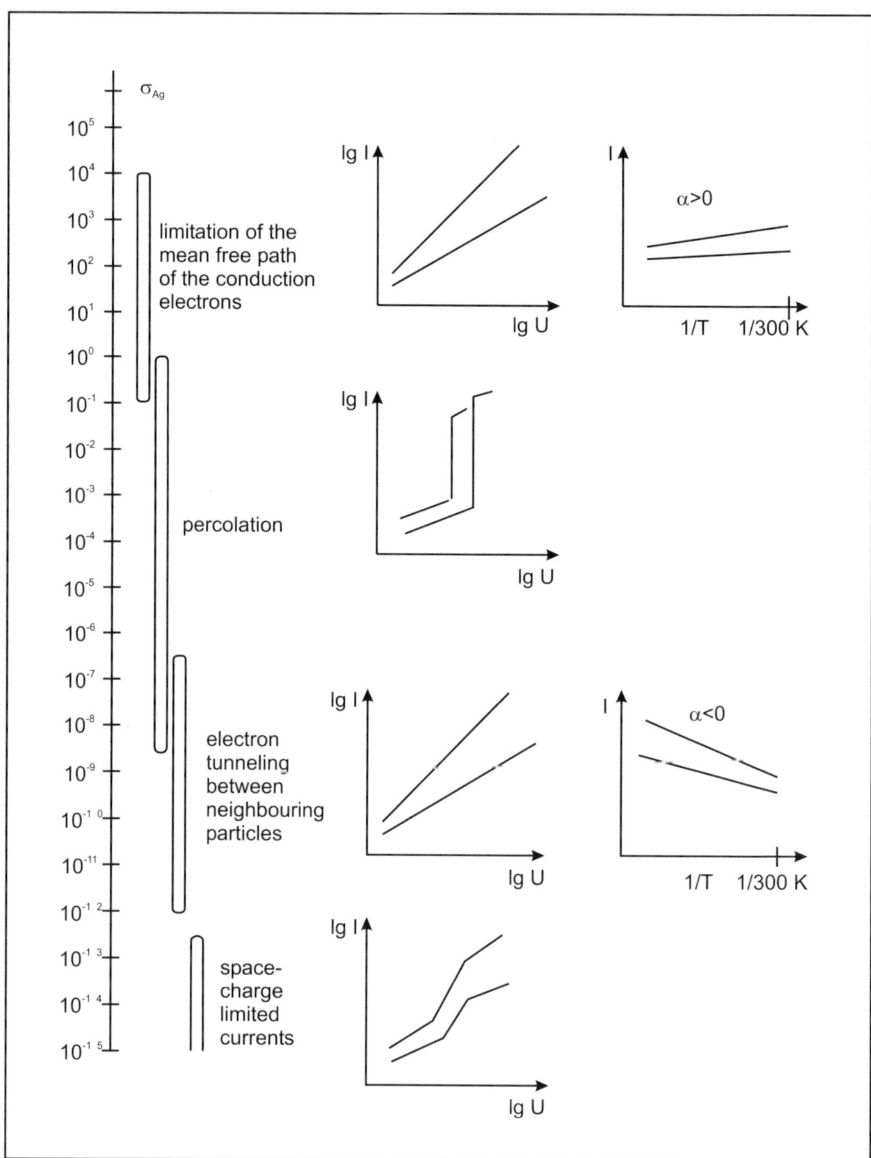

Fig. 5.1. D.c. conductivity of polymer thin films with embedded metal nanoparticles in the metallic, percolation, and dielectric structure ranges

for example. For the particle assemblies discussed in the following, quantum size effects are irrelevant because the majority of the embedded particles are too large.

If space-charge-limited (SCL) charge carrier transport is assumed, it is applicable at very small filling factors ($f \approx 0$) to explain an increase in insulator conductivity caused by the trapping effect of the metal particles.

Effective medium theories can be used to calculate the electrical conductivity of composite materials from bulk values for insulator and metal for all filling factors $0 \leq f \leq 1$. Several effective medium theories [e.g., Bruggeman theory, see (6.24)], also contain a geometry-dependent percolation threshold f_c, and a shift in f_c can be modelled with changing film structure [432]. Effective medium theories, which have been applied to plasma polymer films with embedded silver nanoparticles [433], contain no physical information about charge carrier generation or charge transport. However, they can be used empirically, for example, to make a rough estimate of the conductivity if the filling factor is known, or to determine the influence of different film nanostructures on conductivity if the enlargement for parallel oriented ellipsoidal particles (6.25) and the depolarization factor L are used. In this case (6.25) gives a simple expression for the relationship between percolation threshold and nanostructure.

Limitation of the Mean Free Path. As for thin, continuous metallic films, the film surface and interfaces between metal particles and insulator inclusions in the metallic structure range provide an additional source of conduction electron reflections, and therefore a decrease in the d.c. electrical conductivity σ relative to the d.c. electrical conductivity σ_0 of the bulk metal.

The decrease in the mean free path of conduction electrons λ_e in the bulk metal (for silver $\lambda_{e0} = 57.5$ nm, and for gold $\lambda_{e0} = 40.6$ nm, at 273 K [431]) by reflection of conduction electrons at the film surface is explained by the Fuchs–Sondheimer theory (mean free path theory) in [431, 434, 435]. If the mean free path is known, the film-thickness-dependent d.c. conductivity $\sigma(d)$ can be determined from the d.c. conductivity σ_0 of the bulk metal using

$$\sigma(d) = \sigma_0 \left[1 - \frac{3\lambda_e}{8d} \right] , \quad \text{where} \quad \frac{d}{\lambda_e} \gg 1 , \tag{5.1}$$

and

$$\sigma(d) = \sigma_0 \frac{3d}{4\lambda_e} \left[\ln \frac{\lambda_e}{d} + 0.423 \right] , \quad \text{where} \quad \frac{d}{\lambda_e} \ll 1 . \tag{5.2}$$

Equation (5.1) is more applicable to thin, continuous silver or gold layers ($d \geq 50$ nm) at temperatures $T > 273$ K, because $\lambda_e \ll \lambda_{e0}$ in this range. Equation (5.2) only applies to silver and gold films at low temperatures, because λ_{e0} is substantially higher at low temperatures [431].

The d.c. conductivity increases during thermal treatment ($T \geq 295$ K) of thin metallic films. This is due to a decrease in structural defects developed during film deposition, particularly with regard to grain boundaries between the single crystallites of a polycrystalline metallic film [436]. Once the number of structural defects in the film has been substantially reduced, a temperature coefficient $\alpha > 0$ is measured.

Applications of path length theory are problematic even for thin, continuous metallic films without insulator inclusions, since the exact decrease in the mean free path due to defects is often not known. The d.c. conductivity is also influenced by surface roughness [437] and by the different preparation conditions of the metallic thin films. Path length theory becomes even more complicated when applied to insulating films with embedded metal particles, because the shape and size of insulator inclusions are not considered, and because values of the mean free path determined for thin, continuous metallic films at room temperature (e.g., for gold films $d = 26$ nm, $\lambda_e = 23.6$ nm [438], for copper films $d = 24$ nm, $\lambda_e = 26$ nm [439]) can only be used to make a rough estimate.

Percolation. In the percolation region, which is the transition region between metallic and insulating structure ranges, the d.c. conductivity σ is commonly expressed by [440, 441]

$$\sigma \propto (f - f_c)^\mu . \tag{5.3}$$

The theoretical values are $f_c \approx 0.5$ in a two-dimensional disordered particle assembly and $f_c \approx 0.3$ in a three-dimensional composite system [442, 443]. The conductivity exponent, which depends on the percolation model, varies over $1.2 \leq \mu \leq 1.3$ in the two-dimensional case and $1.7 \leq \mu \leq 2.0$ in the three-dimensional case [441, 444]. It is assumed for modelling with random resistance networks that all current flows through a single particle at $f = f_c$, which provides the active connection (backbone) between the electrodes. An overview of percolation theories and models of electrical conductivity close to the percolation threshold is given in [441, 444].

An experimental investigation of d.c. electrical conductivity close to the percolation threshold requires measurements of non-continuous films, e.g., during the deposition of thin metallic films. To measure discontinuous metallic films or insulator films with embedded metal nanoparticles after deposition, it is necessary to fabricate a large number of samples with a nanostructure close to the percolation threshold. Because very small nanostructural changes during the measurement can lead to significant changes in d.c. conductivity, the experimental determination of d.c. conductivity close to the percolation threshold is very difficult.

However, experimental determinations of the percolation threshold f_c and the coefficient of conductivity μ are very important for characterizing thin discontinuous metallic films or insulator films with embedded metal nanoparticles. The filling factor f_c at the percolation threshold is assumed to be where the d.c. conductivity changes suddenly, or where the temperature coefficient of the d.c. electrical conductivity is $\alpha = 0$. The conductivity exponent can be determined from the slope of the graph $\log(\sigma)$ vs. $\log(f - f_c)$, whereby the theoretical values for disordered networks can occasionally be confirmed.

Percolation thresholds of $0.62 \geq f_c \geq 0.54$ were determined in [445] for co-sputtered Au–SiO$_2$ films, and found to decrease with increasing film thickness ($100 \leq d \leq 1\,000$ nm). The conductivity exponent $1.12 \leq \mu \leq 2.16$ increases with film thickness. This can be interpreted as the transition from two- to three-dimensional percolation. Co-evaporated Co–Al$_2$O$_3$ films have a percolation threshold of $f_c \approx 0.25$ [221]. Other measurements give percolation thresholds of $f_c \approx 0.3$ for co-sputtered Ag–SiO$_2$ films [446] and $f_c \approx 0.22$ for Ag particles in KCl [447, 448]. A percolation threshold was found for evaporated Ag films at an area coverage of $F_A \approx 0.74$ [449]. The experimentally determined values for the exponent of conductivity are $\mu = 1.33$ for evaporated Ag films [450], $\mu = 1.25$ for evaporated Au films [451] and $\mu = 1.27$ for evaporated Ni films [452].

The different values for f_c and μ determined experimentally for similar materials show the influence of the detailed nanostructure on electrical properties of films. It is clear that particle sizes, shape distribution, and metal particle/insulator interface properties are all relevant close to the percolation threshold, and that use of the filling factor and mean particle size does not alone yield an adequate characterization of the nanostructure.

Thermally Activated Tunneling. The electrical conductivity of discontinuous metallic films or insulator–metal composite films in the dielectric structure range will be described in terms of charge carrier transport between two neighboring metal particles by quantum-mechanical tunneling. For discontinuous metallic films it is generally assumed that the tunneling transition occurs via the substrate [453]. For insulator–metal composite films, the tunneling transition is influenced by the surrounding insulator material with dielectric constant $\varepsilon_{\text{ins}} = \varepsilon_{\text{r(ins)}} \cdot \varepsilon_0$.

The concept of thermally activated tunneling, suggested by Neugebauer and Webb [454], considers electron transfer between two initially neutral metal particles. This electron transfer leads to an increase in electrostatic energy. Only single charging of the metal particles is considered (no multiple charging). The charged carrier concentration

$$n = N \exp \frac{-\delta E}{k_B T}$$

is derived from the Boltzmann distribution for carrier density N. The activation energy δE can be determined from the radius R and the spacing B between two neighboring particles (5.4).

Thermally activated tunneling according to Neugebauer and Webb:

$$\delta E = \frac{e^2}{4\pi \varepsilon_{\text{ins}}} \left(\frac{1}{R} - \frac{1}{2R + B} \right), \tag{5.4}$$

where δE is the activation energy, R the radius of the embedded metal particles, B the spacing between two neighboring metal particles, and ε_{ins} the dielectric constant of the surrounding insulator material.

The model for thermally activated tunneling was extended by Swanson [455]. Here, it is assumed that

$$2n = N \left[1 + \exp\left(\frac{-\delta E}{k_{\mathrm{B}}T} \right) \right]^{-1}$$

for the charge carrier concentration, but multiple charging of metal particles is still not considered. Contributions from charge carrier separation (tunneling transition between two neutral particles), charge carrier recombination (tunneling between positively and negatively charged particles), and tunneling transitions between positive and negative and neutral particles are taken into account for charge transport in (5.5).

Thermally activated tunneling according to Swanson:

$$\delta E = \frac{e^2}{4\pi\varepsilon_{\mathrm{ins}}} \frac{1}{R} \left[1 + \frac{R}{B} + \left[\frac{R}{B} \right]^2 + \left[\frac{R}{B} \right]^3 + \left[\frac{R}{B} \right]^4 + 3 \left[\frac{R}{B} \right]^5 \cdots \right]^{-1} .$$

$$(5.5)$$

In the specific case of granular metallic films, which are somewhat similar to metal–ceramic composite films, a charge carrier density

$$n = N \exp \frac{\delta E}{2 k_{\mathrm{B}} T}$$

has been used by Abeles [195] under the assumption that two oppositely charged metal particles are always generated. The activation energy is half the activation energy determined by Neugebauer and Webb using (5.4).

The electrostatic activation energy can be determined as a function of particle size and particle spacing, and therefore as a function of the nanostructure of discontinuous metallic films or insulator–metal composite films, using (5.4) and (5.5). A simple calculation shows that the influence of particle spacing B on the activation energy is only significant when using (5.5). Particles with sizes $R > 5$ nm yield values $\delta E \leq 0.1$ eV for both equations, and the activation energy decreases only slightly with increasing particle size R or with increasing particle spacing B.

The dielectric constant of the insulator plays an important role. $\varepsilon_{\mathrm{ins}} = 1$ for thin metallic films in air or vacuum. If it is assumed that tunneling occurs through the substrate, then the dielectric constant of the substrate material can be used (e.g., for SiO_2, $\varepsilon_{\mathrm{ins}} \approx 3.9$). However, the values $\varepsilon_{\mathrm{ins}} = 6$ and $\varepsilon_{\mathrm{ins}} = 14$ are suggested in [456]. Experimentally determined dielectric constants lie in the range $1.8 \leq \varepsilon_{\mathrm{ins}} \leq 5.2$ [457].

Equations (5.4) and (5.5) are applicable to spherical particles with uniform size and constant spacing. The activation energy for non-spherical particles has been determined in [458]. It is assumed that the activation energy, like the particle size, is also subject to a logarithmic normal distribution [459].

So far, only small applied electrical fields have been considered. Higher fields lead to field-induced charge generation (Fowler–Nordheim field emission), which clearly exceeds thermal charge-carrier generation, and to field-assisted tunneling, which is correlated with a sudden increase in d.c. conductivity [195, 460, 461].

The activation energy can be determined experimentally by measuring the d.c. conductivity at different temperatures. The temperature dependence of the d.c. conductivity σ can generally be interpreted with $\ln \sigma(T) \propto T^{-x}$ and [195, 462, 463]

$$\sigma(T) = \sigma_0 \exp \left[\frac{-A}{k_B T} \right]^x \ . \tag{5.6}$$

If $x = 1$, this yields the simple Arrhenius equation

$$\sigma(T) = \sigma_0 \exp \left[\frac{-\delta E}{k_B T} \right] \ . \tag{5.7}$$

If there is a definite linear dependence in the graphical plot $\ln \sigma(T^{-x})$, the electrostatic activation energy δE can be determined from the slope.

At small fields, $x = 1$ was found for discontinuous gold films above [453] and below [464] room temperature. Experimental values for discontinuous metallic films can also be described with $x = 1/2$ [465]. Otherwise, $x = 1/2$ has been determined for Ni–SiO$_2$ cermet films at temperatures below room temperature [195, 466].

A generally satisfactory agreement between experimental and theoretical values for thermally activated tunneling in discontinuous metallic films or insulator films with embedded metal nanoparticles has not yet been achieved [467]. There are many reasons for this. One is the use of mean values for the particle radius R and interparticle distance B without considering the real nanostructure. In addition, there are influences from the measurement setup and contact effects. Further thermally activated tunneling models for conductivity in discontinuous metallic thin films have been suggested, e.g., the hopping model [468] or space-charge-limited currents for small metal particles [253]. However, none of these provide a more satisfactory description of the experimentally determined results in terms of thermally activated tunneling. Compared with the other theories, the theoretical description of the relationship between electrical properties and nanostructure is rather oversimplified.

5.2 D.C. Conductivity of Plasma Polymer Films with Embedded Metal Particles

5.2.1 Sample Preparation

The d.c. conductivity was determined from the current–voltage relationship for films deposited on thermally oxidized silicon wafers with metal slit electrodes. Up to 16 gold electrodes (2.5×10 mm, thickness ≈ 500 nm) were evaporated with a spacing of 500 μm (slit length l_S) using an electrode mask. Plasma polymer films containing embedded nanoparticles with continuously varying filling factors were deposited on the electrode assembly, so that only half of the electrode area was covered (active slit width b_S) to enable further contact via a conductive silver lacquer or micro-tips. One deposition yielded up to 20 samples with continuously changing metal content, sample 1 having the highest.

Multilayer systems were also investigated, but the plasma polymer film deposited first was much thinner than for the films investigated. The first plasma polymer film separates the metal particles from the substrate material and substrate-assisted tunneling between the metal particles can be excluded. If the particles are in direct contact with the substrate material or if the films have vertically constant particle distribution, substrate-assisted tunneling cannot generally be excluded. Otherwise, the lower plasma polymer films have to be as thin as possible in the thickness region near to the interparticle spacings to avoid insulation of the electrodes.

The d.c. electrical conductivity $\sigma(E)$ for different field strengths E can be determined by means of the current–voltage relationship for the electrode geometry, that is, the slit length l_S and active slit width b_S, and the thickness d of the total multilayer:

$$\sigma(E) = \frac{l_\mathrm{S} I}{b_\mathrm{S} d U} \,, \qquad E = \frac{U}{l_\mathrm{S}} \,. \tag{5.8}$$

Since the film thickness of the metal-particle-containing layer is not generally known in multilayer systems, it was calculated for the entire film thickness d. The actual d.c. conductivity of the particle-containing layer is d/d_c higher than the d.c. conductivity calculated for the multilayer system.

Current–voltage measurements were made with a programmable multimeter under PC control in a current range $I \geq 10^{-14}$ A. The applied voltage was stepped up from $U_\mathrm{min} = 0.01$ V to $U_\mathrm{max} = 50$ V to establish a double-logarithmic plot of the current–voltage behavior. The measurement cycle recorded data from $U_\mathrm{min} \longrightarrow U_\mathrm{max}$, and then from $U_\mathrm{max} \longrightarrow U_\mathrm{min}$.

Since the conducting silver lacquer contacts or micro-tips were applied in atmospheric conditions, it was not possible to detect whether there was any change in d.c. conductivity during the first contact of the samples with the air. No differences were observed between air and vacuum measurements for I–U measurements at room temperature. The determination of the sample current during thermal treatment took place in vacuum ($< 10^{-3}$ Pa). The thermal treatment cycle contains the measurement of the sample current at constant voltage from room temperature ($T_R = 293$ K) to the maximum temperature T_H and back from T_H to T_R. The heating rate was about 5 K min^{-1}. Current–voltage relationships were also determined at different temperatures.

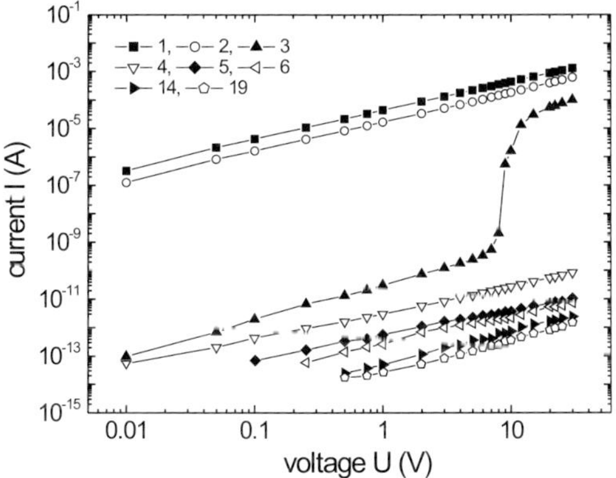

Fig. 5.2. Current voltage characteristic $U_{min} \longrightarrow U_{max}$ for 8 selected samples of polymer thin films with embedded metal nanoparticles

5.2.2 D.C. Conductivity

Figure 5.2 depicts the current–voltage graphs $U_{min} \longrightarrow U_{max}$ for 8 samples of a plasma polymer multilayer system with continuously varying content of embedded silver nanoparticles (film I/20). In the following, the current–voltage relationships are discussed separately for each structural region. Samples 4 to 19 belong to the insulating region. The d.c. conductivities of this sample calculated with (5.8) lie in the region $\sigma = 10^{-6}$–10^{-9} Ω^{-1} cm^{-1}. For the samples 6 to 19, the plots of $U_{min} \longrightarrow U_{max}$ and $U_{max} \longrightarrow U_{min}$ are almost identical and they are not depicted in the diagram. The linear behavior of the current–voltage relationship suggests that thermally activated tunneling acts as a charge transfer process. This can be proved by measurements at different temperatures, as described in Sect. 5.2.3.

In the metallic structural region, samples 1 and 2 also show a linear current–voltage relationship. The d.c. conductivity of these samples is $\sigma = 10^{0}$–10^{-1} Ω^{-1} cm^{-1}. Moreover, for these samples, there is now a substantial difference between the plots of $U_{min} \longrightarrow U_{max}$ and $U_{max} \longrightarrow U_{min}$. Limitation of the mean free path can be assumed to be a conducting mechanism. Once again, this has to proved by measuring the temperature dependence of the sample current.

Figure 5.3 depicts the d.c. conductivity of a plasma polymer film with embedded silver nanoparticles (film I/20). For 19 different sample positions, the d.c. conductivity was calculated using the sample geometry (5.8) and the current values at 1 V. On the basis of TEM investigations [249], the area filling factor was determined for a number of sample positions and is

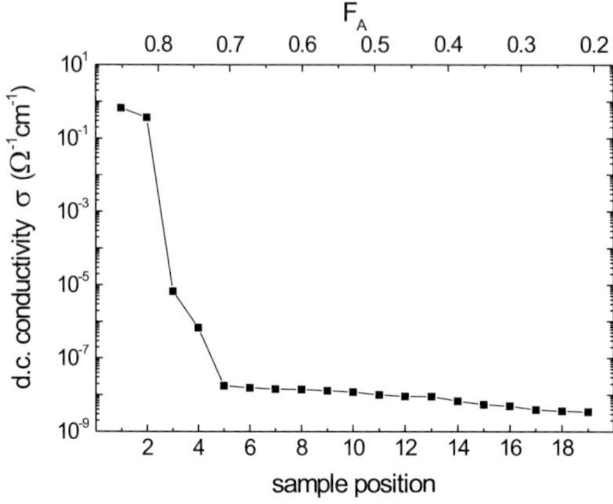

Fig. 5.3. D.c. conductivity σ as a function of sample position and area filling factor F_A for plasma polymer films with embedded silver nanoparticles having continuously varying metal content

also shown in the diagram. Percolation was found at area filling factor $F_A \approx$ 0.75. From considerations given in detail later (Sect. 6.2.4 and Fig. 6.18), the percolation threshold can be assumed to lie in the range $0.32 < f_c < 0.5$. Considering the specific nanostructure of multilayer systems with embedded silver nanoparticles and the fact that only an approximation to the volume filling factor f is possible for multilayer systems, values of f_c can only be given in a large interval.

In [266] and [275], the d.c. conductivity σ is given as a function of the filling factor for a number of plasma polymers made from perfluoropropane and tetrafluoroethylene, respectively, with embedded gold nanoparticles. Because of the approximately constant filling factor, the d.c. conductivity was measured on films deposited sequentially. For all these films, a vertically homogeneous particle distribution was assumed. The percolation threshold was determined as $f_c = 0.38 \pm 0.05$ [266] and $f_c = 0.40$ [275]. Because of the large number of samples in both the metallic and insulating structure ranges, the values for f_c can be assumed to be reliable as determined for films with vertically homogeneous particle distribution.

Because of its nonlinear behaviour, the curve of sample 3 in Fig. 5.2 is of special interest. Obviously, the sample has a nanostructure very close to the percolation threshold. A sudden increase in the sample current occurs at a voltage of about 10 V in the plot $U_{\min} \longrightarrow U_{\max}$, and this leads to a substantially higher sample current $I_{20\,\mathrm{V}} = 10^{-4}$ A compared to the initial value $I_{9\,\mathrm{V}} = 10^{-8}$ A.

Figure 5.4 shows current–voltage plots $U_{\min} \longrightarrow U_{\max}$ and $U_{\max} \longrightarrow U_{\min}$ for another sample (film I/21) which also has a nanostructure close to the percolation threshold. In the $U_{\min} \longrightarrow U_{\max}$ plot, switching occurs at $U_{\mathrm{sw}} = 0.9$ V. The current increases more than six orders of magnitude from $I_{0.8\,\mathrm{V}} = 10^{-12}$ A to $I_{2\,\mathrm{V}} = 10^{-6}$ A. In the $U_{\max} \longrightarrow U_{\min}$ curve, however, no switching back to the lower values was found. This observed switching behavior will be discussed in more detail later. For this purpose, a large

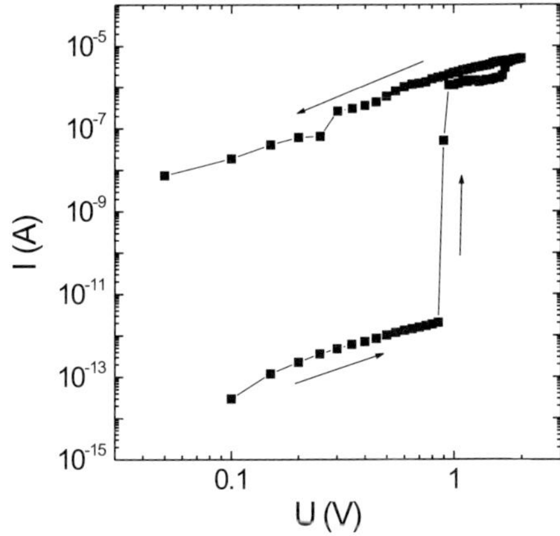

Fig. 5.4. Current–voltage characteristic $U_{\min} \longrightarrow U_{\max}$ and $U_{\max} \longrightarrow U_{\min}$ for a sample with a nanostructure close to the percolation threshold

number of current–voltage characteristics were measured in a well-defined time regime starting with the very first measurement.

Figure 5.5 demonstrates that the sudden change in the sample current and the hysteresis behavior depend on measurements carried out previously. In order to measure a sufficient number of fresh samples, a large number of I/21 films were deposited. The first three measurements $U_{\min} \longrightarrow U_{\max}$ and $U_{\max} \longrightarrow U_{\min}$ from one sample taken amongst these are given. The first measurement $U_{\min} \longrightarrow U_{\max}$ is similar to the curve given in Fig. 5.4. At $U_{\mathrm{sw}} = 1.2$ V the sample current jumps by five orders of magnitude from $I_{1\,\mathrm{V}} = 10^{-12}$ A to $I_{2\,\mathrm{V}} = 10^{-6}$ A. In the current–voltage relationship $U_{\max} \longrightarrow U_{\min}$, there is back-switching at $U_{\mathrm{bsw}} = 0.6$ V from $I_{0.7\,\mathrm{V}} = 10^{-7}$ A to a current value $I_{0.5\,\mathrm{V}} = 10^{-12}$ A which is nearly the same as before switching.

In the second measurement cycle, there is up-switching at $U_{\mathrm{sw}} = 1$ V and down switching $U_{\mathrm{bsw}} = 0.2$ V. Hysteresis has substantially increased. In the third measurement cycle, the switching voltage has increased again ($U_{\mathrm{sw}} = 0.9$ V) and back-switching does not occur. During the three measurements, the initial switching voltage U_{sw} in the $U_{\min} \longrightarrow U_{\max}$ current–voltage relationship decreases from 1.2 V to 0.9 V.

In Fig 5.6, the switching voltage U_{sw} is given as a function of the delay time between subsequent measurement cycles. All four measurements start with fresh samples of film I/21. As the delay time is lowered, the switching voltage decreases rapidly down to a value of $U_{\mathrm{sw}} = 0.2$ V. If there is a long delay between two measurement cycles, the switching voltage stays approximately constant at $U_{\mathrm{sw}} = 1.1$ V.

Obviously, the switching behavior is related to electrostatic charging of the sample during measurements. To avoid these effects, a short circuit was initiated after each measurement cycle. Then, even for a delay time of 60 min, the switching voltage remains at values near $U_{\mathrm{sw}} \approx 1.0$ V. There is no case during the following measurements in which the switching voltage becomes higher than its value for the first switching.

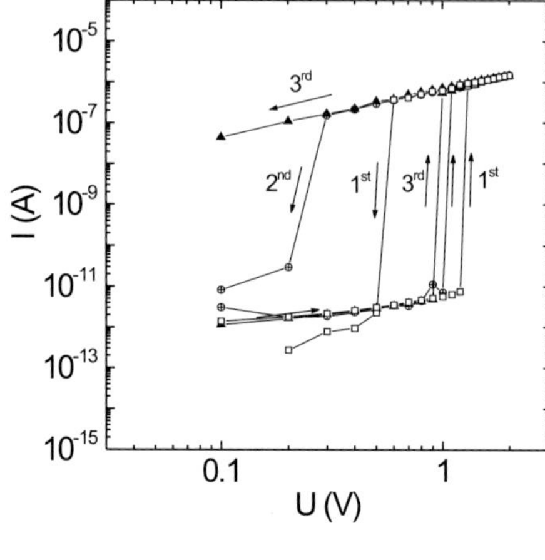

Fig. 5.5. Current–voltage characteristics $U_{\mathrm{min}} \longrightarrow U_{\mathrm{max}}$ and $U_{\mathrm{max}} \longrightarrow U_{\mathrm{min}}$ for the first three measurement cycles

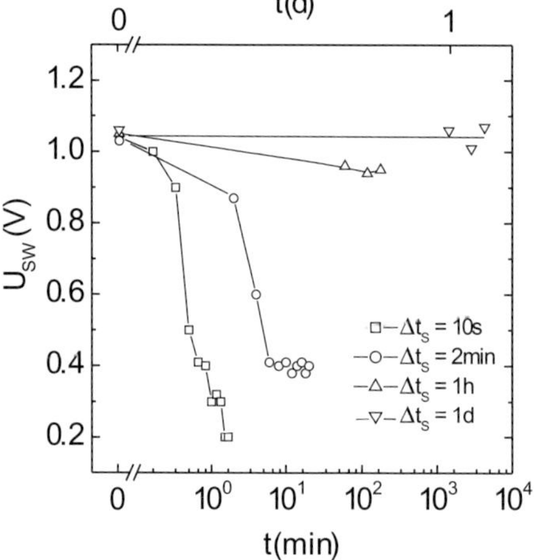

Fig. 5.6. Dependence of the switching voltage U_{sw} on the time delay between measurement cycles. Time delays are 10 s, 2 min, 1 h and 1 d, respectively

For plasma polymer multilayers with embedded silver particles, initial switching voltages in the range $1 < U_{\mathrm{sw}} < 30$ V have been found [433]. Up to now, no significant relationship has been found between the switching voltage and the polymer matrix and deposition properties of the silver nanoparticles. It has been demonstrated [249] that switching can be repeated more than 100 times if sample charging is avoided by applying short circuits.

It should be asked to what extent nanostructural changes could be responsible for this switching effect. In particular, for samples near to the percolation threshold, such nanostructural changes are very likely owing to high local currents along the backbones. In particular, polymer regions separating the very large embedded silver particles are extremely thin. The TEM micrographs in Fig. 3.3 should be mentioned again here. Since charge transport takes place through only a small number of particles which comprise the backbone, high local current densities are possible. For example, local current densities of up to $j = 10^7 \, \mathrm{A\,cm^{-2}}$ can occur for a sample current of $I = 1 \, \mu\mathrm{A}$ and a mean particle cross section of about $1\,000 \, \mathrm{nm^2}$. The local heating and destruction of polymer regions between two neighboring particles which result from these high current densities can lead to an increase in the sample current and initiate the first switching. The described switching behavior during the first three measurements (see Fig. 5.5) suggests that there can be a forming process which slightly changes the nanostructure. However, in principle, nanostructural changes are assumed to be irreversible and cannot explain the reproducibility of the switching voltage (Fig. 5.6). This means that the main reason for switching behaviour must be electronic. Although nanostructural changes cannot generally be neglected, a reversible transition between the two mechanisms for charge carrier generation can be assumed.

Switching can be described as a crossover between two mechanisms for charge carrier generation, i.e., thermal activation and Fowler–Nordheim field emission [195]. An apparent switching behavior at field strength $F \approx 10^5 \, \mathrm{V\,cm^{-1}}$ has also been observed for polymer films with embedded gold nanoparticles made by simultaneous sputtering and plasma polymerization (monomer perfluoropropane, filling factors $f = 0.12$ and $f = 0.28$, respectively) [247]. As for the description given before, this switching was interpreted as a transition between charge carrier generation by thermal activation and by Fowler–Nordheim field emission. Switching behavior has also been observed for plasma polymer films with embedded copper nanoparticles (monomer propane, $f = 0.08$) [248]. In all these earlier investigations, the reproducibility of switching was not demonstrated.

Switching behavior and instabilities in d.c. conductivities have been found for a large number of materials and frequently assigned to microstructural or nanostructural changes. For polyester–silver composite materials (silver particles $D \approx 1 \, \mu\mathrm{m}$) switching was explained by the formation of conductive paths [469]. Further switching and hysteresis phenomena have also been noticed for thin gold films (electroformation) [470, 471] or in a-Si:H films with embedded molybdenum or titanium particles [472]. Changes were observed in the resistance of thin gold films during storage in an oxygen atmosphere [473]. In particular, studies of discontinuous metallic films without embedding in an additional matrix material must be associated with a change in nanostructure due to surface diffusion.

5.2.3 Temperature Dependence of D.C. Conductivity

Measurements were carried out to ascertain the temperature behavior of the current–voltage relationships. The aim was to confirm the supposed conducting mechanism in the metallic and insulating structural regions and to study the way thermally induced nanostructural changes influence electrical behavior. The observed switching, occurring even during a simple determination of the current–voltage characteristic, makes it even harder to determine the temperature dependence of the d.c. conductivity in samples close to the percolation threshold. Since changes in the current–voltage characteristic during measurement can be attributed, for example, to thermally induced changes resulting from high local current densities, it is hard to distinguish between different possible causes of change in the d.c. conductivity.

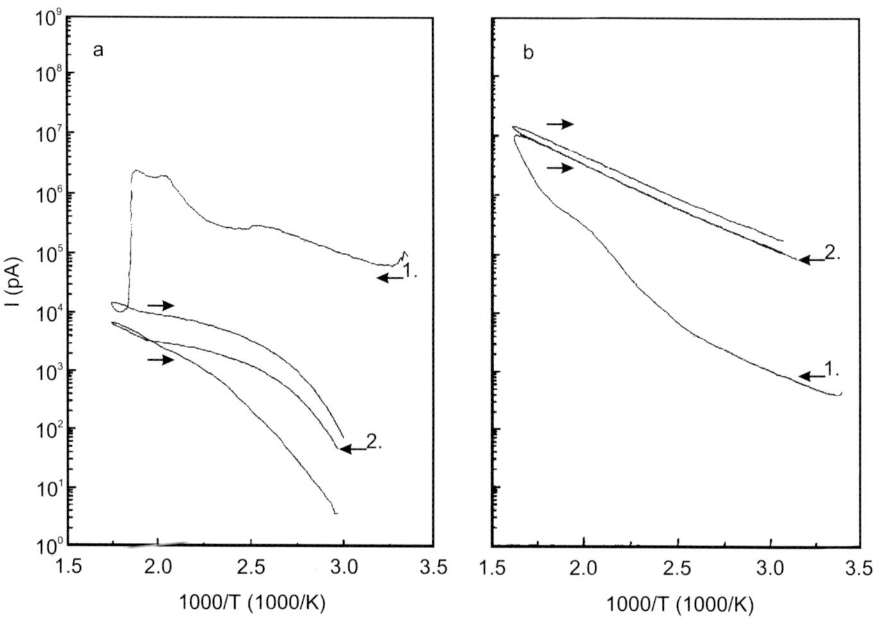

Fig. 5.7. Sample current at constant voltage (**a**) $U = 0.1$ V, (**b**) $U = 1$ V, as a function of reciprocal temperature $1000/T$ for two thermal treatment cycles on two samples

Figure 5.7 shows the dependence of the sample current I on the inverse temperature T^{-1} for two samples from two different plasma polymer multilayer systems (films II/8 and II/9, respectively) with embedded silver nanoparticles, during the first two thermal treatment cycles and at constant sample voltages $U = 0.1$ V (Fig. 5.7a) and $U = 1$ V (Fig. 5.7b). Considering first Fig. 5.7a, it can be observed that initially the sample current increases during the first thermal treatment cycle. Then at a temperature of 535 K, the sample current suddenly decreases by about two orders of magnitude. It can be assumed that changes in the nanostructure occur at this temperature, involving reshaping and coalescence of the silver particles, as described in Sect. 4.1.3.

During cooling, the sample current decreases further, and has a value three orders of magnitude lower than before the thermal treatment. A second treatment cycle leads to a further decrease in the sample current, but without the sudden decrease seen previously. The conductivity before the first thermal treatment cycle was $\sigma \approx 10^{-2}$ $\Omega^{-1}\,\mathrm{cm}^{-1}$. After the second thermal treatment cycle the conductivity was determined as $\sigma \leq 10^{-6}$ $\Omega^{-1}\,\mathrm{cm}^{-1}$. Further cycles up to 580 K did not yield any more significant changes. The dependence of $\log I(T^{-1})$ was not observed to be linear.

A similarly increasing conductivity has been observed for discontinuous gold thin films as the temperature increases up to 510 K. This is followed by a weak decrease in conductivity, until a sudden decrease occurs at 555 K [456]. The break-up of single conductive paths starts at 510 K, and it is assumed that the last path is destroyed at 555 K. This effect has also been observed in thin, discontinuous silver films deposited on PMMA films at 423 K and subsequently cooled. There is an increase in conductivity during thermal treatment, and then a sudden decrease takes place at 340 K.

The other sample in Fig. 5.7b shows completely different behavior of the sample current at constant sample voltage during the two thermal treatment cycles. The sample current I increases throughout the first thermal treatment and at T = 575 K reaches a maximum more than two orders of magnitude higher than the room temperature value. A linear dependence of $\log I(T^{-1})$ is observed during the cooling period. This linear dependence is reproducible in the second and following thermal treatment cycles. The conductivity was determined as $\sigma \approx 10^{-5}$–10^{-6} $\Omega^{-1}\,\mathrm{cm}^{-1}$ before the first thermal treatment cycle, and $\sigma \approx 10^{-3}$ $\Omega^{-1}\,\mathrm{cm}^{-1}$ after the second thermal treatment cycle.

In contrast with the sample presented in Fig. 5.7a, the sample current for the sample in Fig. 5.7b increases continuously during the first thermal treatment. A similar increase in the conductivity has been observed in plasma polymer films with embedded gold nanoparticles in [275].

Conductivity increases can be observed in thin metallic films due to reductions in the number of grain boundaries and crystal lattice defects. The increase in conductivity takes place continuously with the first thermal treatment up to 420 K on thin silver films [474, 475] and 440 K on thin gold films [476]. Up to these temperatures, a reversible linear dependence of the conductivity is observed. The disappearance of grain boundaries, as observed in the electron microscope (see Fig. 4.12), does not happen suddenly.

The different types of behavior of the sample currents for samples in the metallic or percolation regions were mostly observed during the first thermal treatment. A sudden decrease in the sample current at temperatures between 400–500 K can be seen as the result of coalescence and reshaping of metal particles. These abrupt decreases were found in several samples from one plasma polymer thin film with embedded silver nanoparticles at identical temperatures. However, up to now extensive measurements on plasma polymer films made from different monomers and with different embedded metal particles have not yielded any significant dependence of the temperature where nanostructural changes start on the deposition conditions of the plasma polymer matrix.

It is very hard to distinguish between the two processes. The slow and progressive increase in sample current could result from reshaping of metal particles leading to more stable backbones for the current to flow through. Otherwise, a slow, thermally induced reduction in the grain boundaries of the metal particles could also lead to an increase in conductivity.

Irreversible modifications during the first thermal treatment of plasma polymer films with embedded metal nanoparticles are clearly the result of nanostructural changes. The sudden changes in the current–voltage characteristic presented in Fig. 5.7 do not generally recur after the first thermal treatment, and reproducible current–voltage characteristics and current–temperature relationships can then be measured. Changes take place again if further thermal treatments are carried out at higher temperatures.

The nanostructural changes caused by thermal treatments are explained in Sect. 4.2. In addition to these nanostructural changes, which are known from the TEM micrographs, modifications to the plasma polymer between two metal particles can be caused by the current flow, and are not observable in the electron microscope.

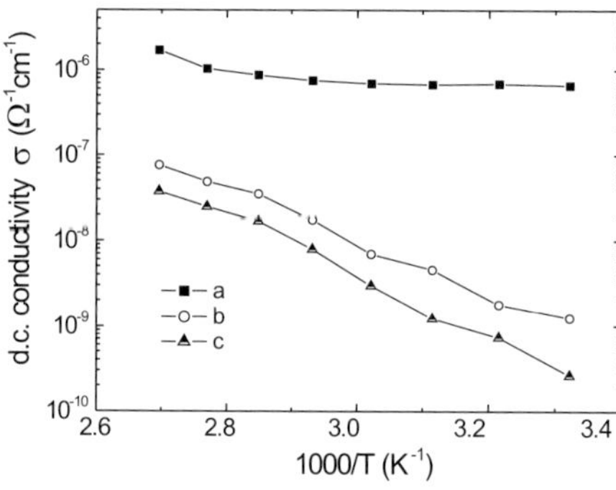

Fig. 5.8. D.c. conductivity for three samples of a plasma polymer multilayer with embedded silver nanoparticles as a function of the inverse temperature during thermal treatment up to 370 K. The constant voltage was 1 V

At this point, where the sample is not further modified after several thermal treatment cycles, it makes sense finally to determine the activation energy. The sample of Fig. 5.7a shows no linear dependence for $\log I(T^{-1})$ or for $\log I(T^{-0.5})$. Figure 5.7b already shows a clear linear dependence for $\log I(T^{-1})$ during the cooling phase of the first thermal treatment cycle. The activation energy δE can thus be determined by means of (5.7). The activation energy for the sample presented in Fig. 5.7b is $\delta E = 0.09$ eV. Larger activation energies of $\delta E = 0.27$ eV, $\delta E = 0.28$ eV and $\delta E = 0.38$ eV have been determined for three further samples of this multilayer system with embedded silver nanoparticles. In other measurements of temperature dependent d.c. conductivity (Fig. 5.8) on a plasma polymer multilayer with embedded silver nanoparticles (film I/21), activation energies of $\delta E = 0.57$ eV (sample b) and $\delta E = 0.62$ eV (sample c) were found, respectively.

Attempts were made to draw conclusions about the particle radius R and particle spacing B from the activation energy values, using the theoretical equations (5.4) due to Neugebauer and Webb and (5.5) due to Swanson. Particle radii of $3 \leq R \leq 5$ nm and $2 \leq R \leq 6$ nm were found from (5.4) and (5.5), respectively, for an activation energy of $\delta E \approx 0.1$ eV, depending on the particle spacing B. These values are too low for the mean particle sizes determined from TEM micrographs for comparable samples. Efforts to match particle sizes determined from (5.4) and (5.5) to those from TEM micrographs, assuming different values of ε_r, do not provide an adequate solution [456].

Experimentally determined activation energies for thin, discontinuous metallic films are:

- $\delta E = 0.04$–0.31 eV for copper films, depending on the effective film thickness and particle size [477],
- $\delta E = 0.02$–0.15 eV $(T \leq T_R)$ [350], $\delta E = 0.02$–0.08 eV $(T \leq T_R)$ [464] and $\delta E = 0.04$–0.09 eV $(T \geq T_R)$ [453], all for discontinuous gold films,
- $\delta E = 0.01$–0.03 eV $(T \leq T_R)$ [478] for copper particles in SiO_2 films.

Attempts to draw conclusions about particle sizes from (5.4) and (5.5) still yield lower particle sizes in these cases than those found by other experimental methods.

For 19 samples from a plasma polymer multilayer with embedded silver particles (film I/20), Fig. 5.9 depicts the sample current at a constant voltage of 1 V as a function of the sample position before and after thermal treatment up to 400 K. Owing to nanostructural changes, values for the area filling factor cannot be given. As explained in Chap. 4.2, the volume filling factor remains constant during thermal treatment whilst the area filling factor changes. For all sample numbers larger than four, the sample current decreases after thermal treatment. As a result of nanostructural changes, the conductivity decreases by approximately one order of magnitude. The percolation threshold is still found close to sample 3.

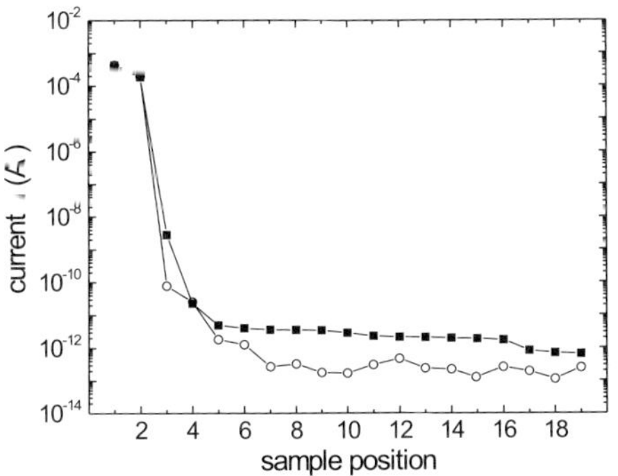

Fig. 5.9. Sample current as a function of sample position for a plasma polymer film with embedded silver nanoparticles having continuously varying metal content before and after thermal treatment up to 400 K

In metal-containing films with a vertical homogeneous particle distribution, the thermally generated nanostructural changes of embedded nanoparticles yield a shift in the percolation threshold f_c. Figure 5.10 shows the specific resistance $\rho(f) = \sigma^{-1}(f)$ in $\Omega\,\mathrm{cm}$ for plasma polymer films made from tetrafluoroethylene with embedded gold nanoparticles before and after thermal treatment up to 475 K [256]. The percolation threshold is shifted from $f_c = 0.38 \pm 0.03$ before thermal treatment to $f_c = 0.42 \pm 0.03$ after thermal treatment.

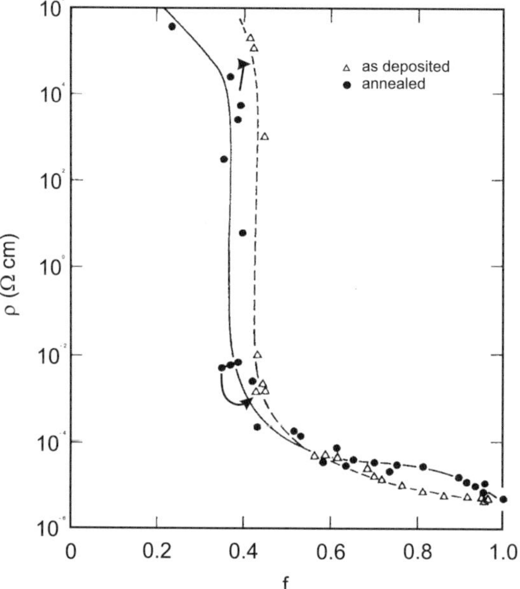

Fig. 5.10. Specific resistance $\rho(f)$ of plasma polymer gold composite films before ($\bullet$) and after ($\triangle$) thermal treatment up to 475 K [256]

Figure 5.10 demonstrates the dependence of the percolation threshold f_c on the nanostructure of nanoparticle-containing films. It is clear that there are samples which have high conductivity before thermal treatment, and belong to the metallic structure range, but which have only a low conductivity after thermal treatment, so that they may then belong to the dielectric structure range. The difference in conductivity between thermally untreated and thermally treated samples can be up to ten orders of magnitude. Combined with the possibility of limiting nanostructural changes on areas irradiated by electron or laser beams, these large changes in d.c. conductivity give polymer thin films with embedded nanoparticles a considerable potential as regards achieving functional nanostructured materials.

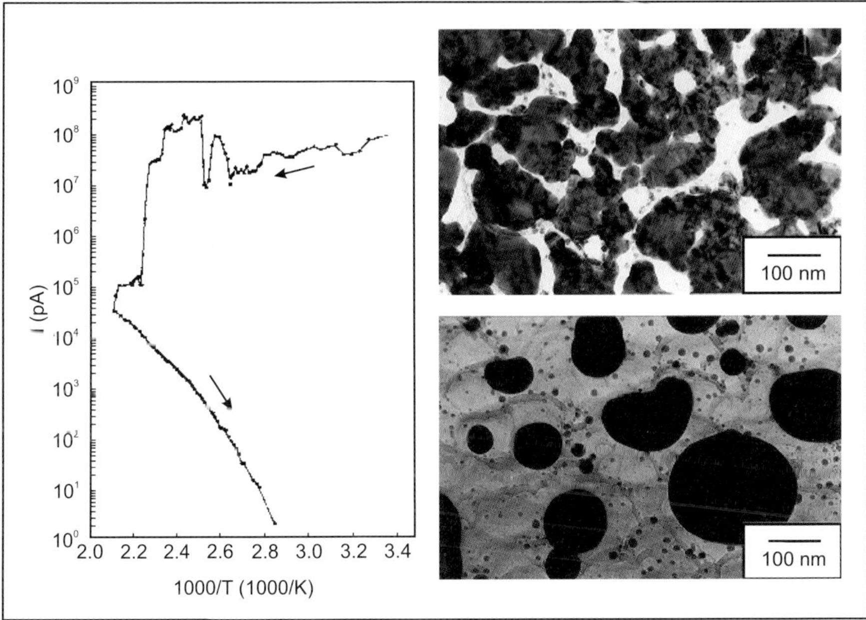

Fig. 5.11. D.c. conductivity and nanostructure for two comparable samples of a plasma polymer multilayer with embedded silver nanoparticles before and after thermal treatment. *Left*: sample current I [pA] at $U = 1$ V during the first thermal treatment cycle up to 470 K. *Right*: TEM micrograph before (*top*) and after (*bottom*) thermal treatment

5.2.4 Nanostructure and D.C. Conductivity

In summary, for a given sample, it should be possible to explain the electrical transport properties of polymer thin films with embedded metal nanoparticles in terms of the film structure. Figure 5.11 shows the behavior of the temperature dependent sample current I [pA] at a constant voltage of $U = 1$ V during the first thermal treatment up to 470 K for a sample of a plasma polymer multilayer system with embedded silver nanoparticles (film I/20). Before thermal treatment the sample has a d.c. conductivity significantly higher than the percolation threshold f_c with $\sigma = 10^0$–10^{-1} Ω^{-1} cm^{-1}.

First, the sample current decreases during thermal treatment. Therefore the temperature coefficient of electrical resistance (TCR) α is positive. After some instabilities with brief increases in the sample current, there is a sudden decrease in the sample current of about four orders of magnitude caused by the onset of coalescence or reshaping at 440 K. The sample current continuously decreases as it cools down, and a negative TCR α is observed.

The d.c. conductivity is $\sigma \leq 10^{-8}$ Ω^{-1} cm^{-1} after thermal treatment and has changed by about eight orders of magnitude compared to the original value. While the sample belongs to the metallic structure range before thermal treatment, it belongs to the dielectric structure range after thermal treatment. Since material loss can be neglected, the percolation threshold f_c has changed to higher filling factors during thermal treatment (see Fig. 5.7). While the d.c. conductivity before thermal treatment can be explained using the mean free path theory, thermally activated tunneling between two neighboring metal particles is assumed after thermal treatment.

The nanostructural changes which are responsible for the decrease in d.c. conductivity are shown in the TEM micrographs of Fig. 5.11. This sample with comparable metal content was taken from the same multilayer system and has been thermally treated in the electron microscope. It can be seen that single, isolated metal particles are generated by the thermal treatment from previously interconnected metal particles.

The electrical transport properties of polymer films with embedded metal nanoparticles as presented result mainly from the granular nanostructure of the metal-containing layers. The materials of the basic components play a minor role. Because of the special nanostructure of samples near the percolation threshold, hysteresis effects occur which have nanostructural origins. However, a simple thermal treatment up to 470 K is sufficient to stabilize electrical properties. Furthermore, switching behaviour was observed in the conductivity. This was attributed to a change in the electronic transport mechanism from thermal activation to Fowler–Nordheim field emission.

The comparison between theoretical models for conductivity mechanisms and experimentally determined results is unequivocal and makes it clear that there are many unanswered questions, as for thin metallic films. There are sure to be other, perhaps more appropriate materials available for experimental investigations to seek new physical formulations of charge carrier generation and transport in granular systems. Nevertheless, the simple ability to make nanostructural modifications and control the resulting changes, e.g., in d.c. conductivity, offers great potential for various applications. Moreover, polymer films with nanoparticles are suited as single-electron tunneling devices [181, 183].

6. Nanostructure and Optical Properties

6.1 Optical Properties of Polymer Films with Embedded Metal Particles

6.1.1 Optical Plasma Resonance Absorption of Embedded Metal Nanoparticles

The optical properties of metal nanoparticles differ substantially from the optical properties of bulk materials. Under the influence of an electric field, there is a plasmon excitation of electrons at the particle surface. This resonance which takes place at a certain energy of the incident light results in optical absorption, the so-called plasma resonance absorption. The phenomenon of plasma resonance absorption will be described briefly to give a basis for understanding the present chapter. Comprehensive reviews of the optical properties of metal nanoparticles are given in [5–7, 479].

The complex dielectric function of the bulk metal $\hat{\varepsilon}_{\mathrm{me}}(\tilde{\nu})$ is determined by the susceptiblity of the free electrons χ^{DS} and the susceptibility of the interband transition in lower electron energies χ^{IB} :

$$\hat{\varepsilon}_{\mathrm{me}}(\tilde{\nu}) = 1 + \chi^{\mathrm{DS}} + \chi^{\mathrm{IB}} \ . \tag{6.1}$$

If only free electrons are considered, the Drude–Sommerfeld theory [6, 480] gives the dielectric function $\hat{\varepsilon}_{\mathrm{me}}(\tilde{\nu})$ as

$$\hat{\varepsilon}_{\mathrm{me}}(\tilde{\nu}) = 1 - \frac{(\tilde{\nu}_{\mathrm{p}}^{\mathrm{DS}})^2}{\tilde{\nu}^2 + i\tilde{\nu}_{\mathrm{c}}^{\mathrm{DS}}\tilde{\nu}} \ . \tag{6.2}$$

Here the plasma frequency $\tilde{\nu}_{\mathrm{p}}^{\mathrm{DS}}$ and the collision frequency $\tilde{\nu}_{\mathrm{c}}^{\mathrm{DS}}$ are constants of the material given by

$$\tilde{\nu}_{\mathrm{p}}^{\mathrm{DS}} = \sqrt{\frac{Nc^2}{\pi m_{\mathrm{e}} c^2}} \quad \text{and} \quad \nu_{\mathrm{c}}^{\mathrm{DS}} = \frac{v_{\mathrm{F}}}{2\lambda_{\mathrm{e}} c} \ , \tag{6.3}$$

where N, m_{e}, and λ_{e} are the number, mass, and mean free path of the electrons, e is the elementary charge, c the speed of light, and v_{F} the Fermi velocity. The plasma frequency $\tilde{\nu}_{\mathrm{p}}^{\mathrm{DS}}$ is defined by $\varepsilon_{\mathrm{me}}'(\tilde{\nu}_{\mathrm{p}}) = 0$ and its values are $\tilde{\nu}_{\mathrm{p}}^{\mathrm{DS}} = 74\,000$ cm^{-1} (9.2 eV) for silver and $\tilde{\nu}_{\mathrm{p}}^{\mathrm{DS}} = 73\,000$ cm^{-1} for gold.

The collision frequency is given by $\tilde{\nu}_{\mathrm{c}}^{\mathrm{DS}} = 145$ cm^{-1} (17.8 meV) for silver and $\tilde{\nu}_{\mathrm{c}}^{\mathrm{DS}} = 215$ cm^{-1} (26.6 meV) for gold [481].

For silver and gold, the interband transition contributes significantly to optical extinction which results in a substantial red shift of the plasma frequency $\tilde{\nu}_{\mathrm{p}}$. The values determined experimentally for the so-called bulk plasmon are $\tilde{\nu}_{\mathrm{p}} \approx 31\,500$ cm^{-1} (≈ 3.9 eV) for silver and $\tilde{\nu}_{\mathrm{p}} \approx 19\,300$ cm^{-1} (≈ 2.4 eV) for gold [6] [see $\hat{\varepsilon}_{\mathrm{Ag}}(\tilde{\nu})$, Fig. 6.13]. For thin but still closed metal films, the result is a transmission maximum and a reflection minimum at $\tilde{\nu}_{\mathrm{p}}$.

For metal particles with radii substantially lower than the wavelength of the incident light, polarization of the conducting electrons takes place at the particle surface under the influence of the electric field. Because of the dipole interaction, small spherical particles with radius R embedded in an insulating matrix have polarization α given by

$$\alpha(\tilde{\nu}) = 4\pi\varepsilon_0 R^3 \, \frac{\hat{\varepsilon}_{\mathrm{me}}(\tilde{\nu}) - \hat{\varepsilon}_{\mathrm{po}}(\tilde{\nu})}{\hat{\varepsilon}_{\mathrm{me}}(\tilde{\nu}) + 2\hat{\varepsilon}_{\mathrm{po}}(\tilde{\nu})} \; . \tag{6.4}$$

For parallel oriented ellipsoidal particles (half axes R_A, R_B, R_C, parallel orientation of the semi-major axis R_A), the polarization α is given by

$$\alpha(\tilde{\nu}) = \frac{4\pi}{3}\varepsilon_0 R_A R_B R_C \frac{\hat{\varepsilon}_{\mathrm{me}}(\tilde{\nu}) - \hat{\varepsilon}_{\mathrm{po}}(\tilde{\nu})}{\hat{\varepsilon}_{\mathrm{po}}(\tilde{\nu}) + L[\hat{\varepsilon}_{\mathrm{me}}(\tilde{\nu}) - \hat{\varepsilon}_{\mathrm{po}}(\tilde{\nu})]} \; , \tag{6.5}$$

where L is a depolarization factor with $0 \leq L \leq 1$ [482, 483]. For ellipsoids which are randomly oriented with respect to the electric field vector, three depolarization factors L_1, L_2, L_3 with $\sum_i L_i = 1$ are required [482, 483]:

$$\alpha(\tilde{\nu}) = \frac{4\pi}{3}\varepsilon_0 R_A R_B R_C \sum_{i=1}^{3} \frac{\hat{\varepsilon}_{\mathrm{me}}(\tilde{\nu}) - \hat{\varepsilon}_{\mathrm{po}}(\tilde{\nu})}{\hat{\varepsilon}_{\mathrm{po}}(\tilde{\nu}) + L_i[\hat{\varepsilon}_{\mathrm{me}}(\tilde{\nu}) - \hat{\varepsilon}_{\mathrm{po}}(\tilde{\nu})]} \; . \tag{6.6}$$

For spherical metal particles, the polarization reaches its maximum at

$$\hat{\varepsilon}_{\mathrm{me}}(\tilde{\nu}) + 2\hat{\varepsilon}_{\mathrm{po}}(\tilde{\nu}) = 0 \; , \tag{6.7}$$

whilst for parallel oriented ellipsoids, it reaches a maximum at

$$L\hat{\varepsilon}_{\mathrm{me}}(\tilde{\nu}) + (1 - L)\hat{\varepsilon}_{\mathrm{po}}(\tilde{\nu}) = 0 \; . \tag{6.8}$$

Equations (6.7) and (6.8) are resonance conditions for the surface plasmon of a metal particle.

In optical spectra, excitation of the surface plasmon can be found as an absorption maximum or a transmission minimum (plasma resonance absorption), with spectral position $\tilde{\nu}_{\mathrm{r}}$, half-width Γ and relative intensity I_{m}. Depending on the dielectric function of the bulk metals, the plasma resonance absorption of silver and gold nanoparticles is found in the visible and near-infrared spectral region. If the particle geometry is known and the dielectric functions of the metal $\hat{\varepsilon}_{\mathrm{me}}(\tilde{\nu})$ and the polymer $\hat{\varepsilon}_{\mathrm{po}}(\tilde{\nu})$ are also known, the spectral position of the plasma resonance absorption can be determined for spherical particles using (6.7) and for ellipsoidal particles using (6.8), respectively.

From the resonance conditions (6.7) and (6.8), the spectral position of the plasma resonance absorption can be calculated. In order to calculate optical spectra which can be compared with experimental spectra, two approaches are useful. First, optical extinction spectra can be calculated using the particle polarization α and compared with experimental extinction spectra (exact route). The other approach (the statistical route) is to introduce an effective dielectric function $\hat{\varepsilon}(\tilde{\nu})$ which contains the optical properties of the matrix and the embedded nanoparticles. The effective dielectric function can be calculated using a so-called effective medium theory. A simple way to derive effective medium theories is to combine the polarization equations (6.4), (6.5), and (6.6) with the Clausius–Mossotti equation (6.42) which describes the electric field inside a dielectric bulk material. The result is the Maxwell–Garnett effective medium theories for spherical (6.21), parallel oriented (6.22), and randomly oriented ellipsoidal nanoparticles (6.23), all introduced in Sect. 6.2.3. From the dielectric functions, transmission and reflection spectra can be calculated using the Fresnel formulas. Generally, effective medium theories make it possible to consider additional nanostructural information about particle assemblies such as the particle size and shape distribution and interface layers between nanoparticles and matrix, for example. For thin films, the film thickness can be taken into account.

Up to now, only dipole excitations have been considered in (6.4), (6.5) and (6.6) for the polarization of the metal nanoparticles. This is a useful approximation if the particles are substantially smaller than the wavelength of the incident light (quasistatic approximation or Rayleigh limit). For larger particles, multipole excitations must also be considered. The Mie theory [5, 484] gives a complete description of the electromagnetic excitation of a metal particle without size restrictions. The optical extinction is a sum over the partial waves of the electric and magnetic fields for dipole, quadrupole and higher excitations. All the above-mentioned statements concerning plasma resonance absorption for spherical particles are included in the Mie theory as a special solution for dipole excitation. Application of Mie theory to particle assemblies involves a substantial computing effort. A detailed description and calculations are given in [6, 485]. The main disadvantage of the Mie theory is that rigorous solutions are only given for spherical particles and a few other geometrical bodies such as spheroids [327] or infinite cylinders [487, 488]. Real particle assemblies with substantial deviations in the particle size and shape can only be considered as coagulated particles

The optical properties of very small nanoparticles are also influenced by quantum size effects, first noticed by Kubo [8, 10]. The mean particle size of the embedded particles described in this book is not less than $\bar{D} = 5$ nm. The particle size for which quantum size effects become important is $D \leq 3$ nm or fewer than 1 000 atoms for a silver particle [489]. It can thus be assumed that quantum size effects do not substantially influence optical properties of the embedded metal particles described in the following.

6.1.2 Experimental Optical Properties of as-Deposited Films

In the following, only metal-nanoparticle-containing multilayer systems with composite film thickness $d_\mathrm{c} \leq 100$ nm and continuously varying filling factor were investigated. One advantage of such films is the possibility of making direct optical measurements of optical transmission spectra without further sample preparation and determining the spectral position, half-width and intensity of the plasma resonance absorption simply from these transmission spectra. In contrast to films with constant filling factor, the continuously varying filling makes it possible to carry out measurements at various positions of the sample with constant polymer matrix properties but with different particle size and shape distributions. Spectra were taken by simply moving the film in constant steps through the beam of the spectrophotometer. For each film, various spectra were measured, e.g. in 2 mm steps, for different metal particle size and shape distributions.

Because of the relative large sample area $(2\ \mathrm{cm}^2)$, reflection measurements using an Ulbricht sphere were only carried out on samples from plasma polymer films with embedded silver particles with constant filling factor. The measured values were corrected using different reflection standards.

Using UV–visible–NIR transmission spectra in the wave number region $50\,000 \geq \tilde{\nu} \geq 4\,000\ \mathrm{cm}^{-1}$, the following discussion of optical plasma resonance absorption will be based on the spectral position $\tilde{\nu}_\mathrm{r}$, half-width Γ and relative intensity $I_\mathrm{m} = T_0/T_\mathrm{m} - 1$, where T_0 is the transmittance of the polymer film at $\tilde{\nu}_\mathrm{r}$ and T_m is the transmittance of the metal-nanoparticle-containing multilayer polymer films at $\tilde{\nu}_\mathrm{r}$.

To begin with, the influence of various plasma polymer matrix parameters will be investigated. Optical properties of the plasma polymer depend on the power density of plasma polymerization (Fig. 2.5). Figure 6.1 shows the transmission spectra of two plasma polymer thin films with embedded silver particles. As can be seen in Fig. 3.8, the mean silver particle size decreases and the mean shape factor increases continuously from spectrum (a) to spectrum (f). The left-hand film (film I/12) was deposited by plasma polymerization of benzene at low power density ($p = 0.03\ \mathrm{W\,cm}^{-2}$) and the thickness of the multilayer system was $d = 80$ nm. In the right-hand film (film I/13), the power density for plasma polymerization was $p = 0.85\ \mathrm{W\,cm}^{-2}$ and the film thickness $d = 150$ nm. The different slopes for transmission at $\tilde{\nu} \geq 40\,000\ \mathrm{cm}^{-1}$ depend on the polymer structure and were discussed in Sect. 2.1.3.

The table in Fig. 6.1 gives the spectral position $\tilde{\nu}_\mathrm{r}\ [\mathrm{cm}^{-1}]$, the half-width $\Gamma\ [\mathrm{cm}^{-1}]$ and the relative intensity I_m of the transmission minima caused by plasma resonance absorption of the silver nanoparticles. With decreasing mean particle size and increasing shape factor, the plasma resonance absorption at $\tilde{\nu}_\mathrm{r}$ shifts to higher wave numbers and its relative intensity decreases. In addition, the half-width decreases. This behavior was found for both films. For comparable positions of the transmission minima, a lower half-width and relative intensity were found for plasma resonance absorption on the film

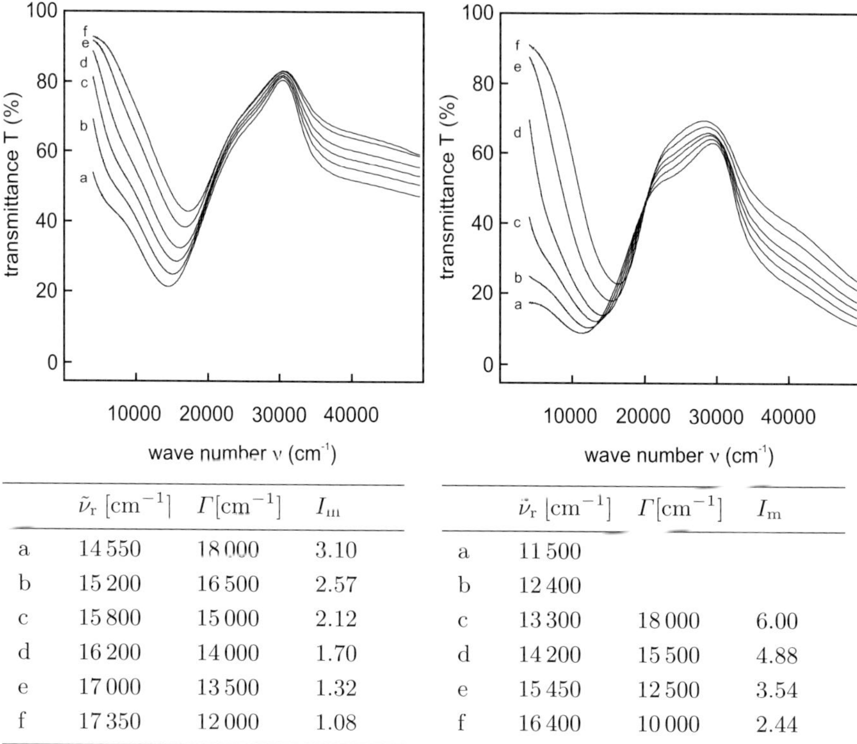

	$\tilde{\nu}_{\mathrm{r}}\,[\mathrm{cm}^{-1}]$	$\Gamma\,[\mathrm{cm}^{-1}]$	I_{m}
a	14 550	18 000	3.10
b	15 200	16 500	2.57
c	15 800	15 000	2.12
d	16 200	14 000	1.70
e	17 000	13 500	1.32
f	17 350	12 000	1.08

	$\tilde{\nu}_{\mathrm{r}}\,[\mathrm{cm}^{-1}]$	$\Gamma\,[\mathrm{cm}^{-1}]$	I_{m}
a	11 500		
b	12 400		
c	13 300	18 000	6.00
d	14 200	15 500	4.88
e	15 450	12 500	3.54
f	16 400	10 000	2.44

Fig. 6.1. Experimental transmission spectra of two multilayer systems with embedded silver nanoparticles having different polymer matrix properties. From spectrum (a) to spectrum (f), the particle size decreases and the particle shape factor increases continuously

deposited at higher power density. On this film, a point of constant transmission was also found. At this point, where $\tilde{\nu} = 20\,500$ cm^{-1} and $T = 46\%$, the optical transmittance does not depend on particle size and shape.

The constant transmission point was found mainly in multilayer systems deposited by alternating plasma polymerization and metal evaporation and also sometimes in films deposited by simultaneous plasma polymerization and metal evaporation. The spectral position of the constant transmission point lies in the range $20\,000 \leq \tilde{\nu} < 23\,000$ cm^{-1} and the transmittance in the range $40 \leq T \leq 60\%$. The occurrence of this point is discussed after the optical calculations in Sect. 6.2.4.

Figure 6.2 shows the transmission spectra of two multilayer films with embedded silver nanoparticles (films I/12 and I/13) deposited under the same conditions as in Fig. 6.1 but with substantially larger silver particles. On the left, in spectra (b) to (f), the transmission minima at $\tilde{\nu}_{\mathrm{r}} = 14\,500$–$12\,000$ cm^{-1} are still visible. With increasing particle size, a transmission maximum occurs

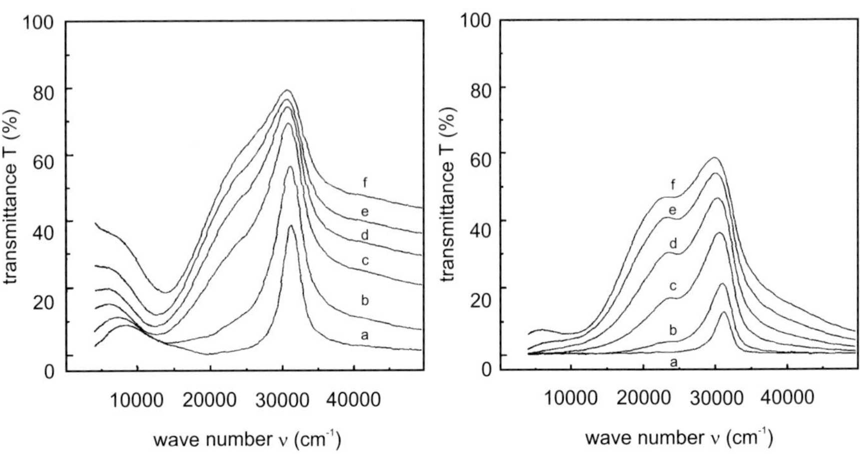

Fig. 6.2. UV–visible–NIR transmission spectra for two multilayer films with embedded silver nanoparticles having different polymer matrix properties. From spectrum (a) to spectrum (f), particle size decreases and particle shape factor increases continuously

at $32\,000$ cm^{-1}. These transmission maxima are found at the same position as the resonance condition for the bulk plasmon of silver at $\varepsilon'(\tilde{\nu}) = 0$, which is the case at $31\,500$ cm^{-1} or 3.9 eV [490]. TEM investigations (Fig. 3.3) demonstrated that the large embedded nanoparticles form a nearly closed thin metal film inside the plasma polymer matrix. Furthermore, in the transmission spectra on the right of Fig. 6.2, e.g., spectra (c), (d), (e) and (f), a second weak transmission minimum was found. Two explanations can be given for this second extinction maximum. Firstly, the multipole polarizability of large silver nanoparticles is observed to be similar to that of silver particles deposited by microlithographic methods [491]. Secondly, particles can be described as long ellipsoids and the plasma resonance absorption splits into two absorptions along major and minor axes of these ellipsoidal particles.

The spectral position and half-width of the plasma resonance absorption are no longer influenced by the polymer matrix once a certain power density is attained during plasma polymerization (monomer benzene, $p \approx 0.15$ W cm^{-2}; monomer HMDSN, $p \approx 0.4$ W cm^{-2}). This is illustrated in Fig. 6.3. Six spectra were taken from various plasma polymer films with embedded silver nanoparticles deposited at six different plasma polymerization power densities. The deposition time ($t_{\mathrm{pp}} = 5$ min) was the same for all the multilayer systems, but because of the different growing rates, they have different thicknesses. All spectra show the same spectral position for plasma resonance absorption, at $\tilde{\nu}_{\mathrm{r}} = 21\,000$ cm^{-1}. For spectra A, B, C, and D, the plasma resonance absorption has the same half-width Γ. Starting at $\tilde{\nu} \geq 25\,000$ cm^{-1}, optical transmission is determined by different optical properties of the plasma polymer.

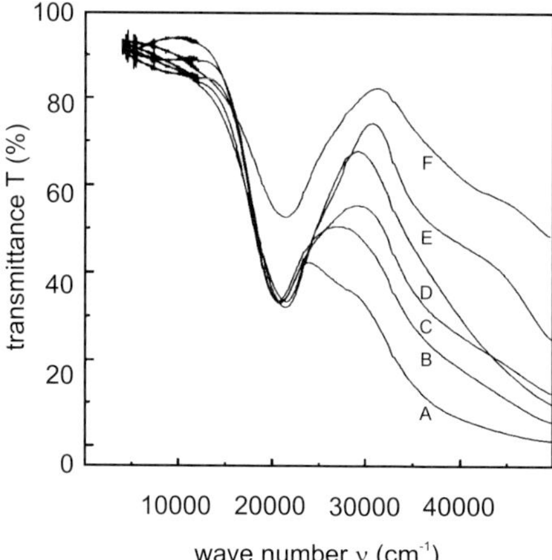

Spectrum	Power density p [W cm^{-2}]	Thickness d [nm]
A	5.0	350
B	3.0	340
C	1.9	390
D	1.0	480
E	0.4	500
F	0.15	510

Fig. 6.3. UV–visible–NIR transmission spectra for six multilayers with embedded silver nanoparticles. Samples were selected with the same spectral position of the plasma resonance at $\tilde{\nu}_r = 21\,000$ cm^{-1}

The dielectric functions of plasma polymer films with embedded silver nanoparticles were calculated (Fig. 6.4) using optical transmission and reflection spectra (Fig. 6.5). The mean particle size decreases and the mean shape factor increases through curves (a) to (c). Curve (d) was calculated for a metal-free plasma polymer thin film using the data from Fig. 2.6. The formation of multilayer systems was considered in the calculation [146]. With increasing metal particle size and decreasing mean shape factor, there is a red shift in the spectral position, an increase in the half-width and an increase in the relative intensity of the plasma resonance absorption visible as a maximum in the real part of the dielectric function. Together with transmission spectra from Fig. 6.1, the dielectric function given in Fig. 6.4 will be discussed further in Sect. 6.2.4, which describes the results of optical calculations.

The transmission spectra given in Figs. 6.1 and 6.2 are representative of a large majority of plasma polymer multilayers with embedded metal particles prepared by simultaneous or alternating plasma polymerization and metal evaporation. Varying the thicknesses d_{p1} and d_{p2} of embedding films does not substantially influence spectra. Because the plasma polymer has different dielectric functions $\hat{\varepsilon}_{po}(\tilde{\nu})$ for various power densities (Fig. 2.7), this plasma polymerization parameter has the main influence on optical transmission spectra. The occurrence of a constant transmission point requires a minimum power density of plasma polymer deposition, e.g., for benzene monomers $p \approx 0.15$ W cm^{-2}. Up to now, the spectral position of the constant transmission point has not been found to depend significantly on the deposition power density. The constant transmission point was not observed for silver particles deposited directly on quartz glass without a plasma polymer matrix.

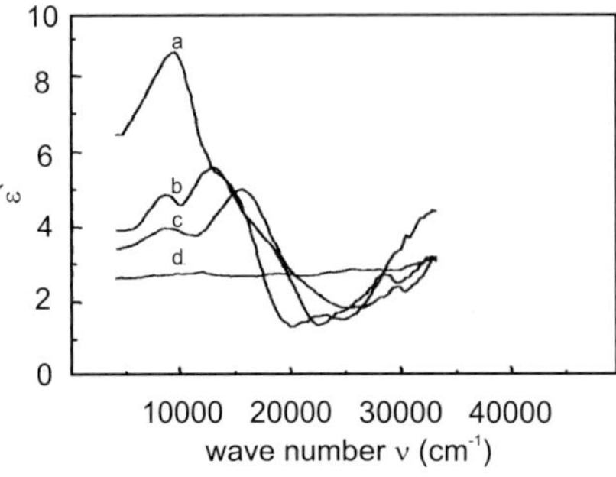
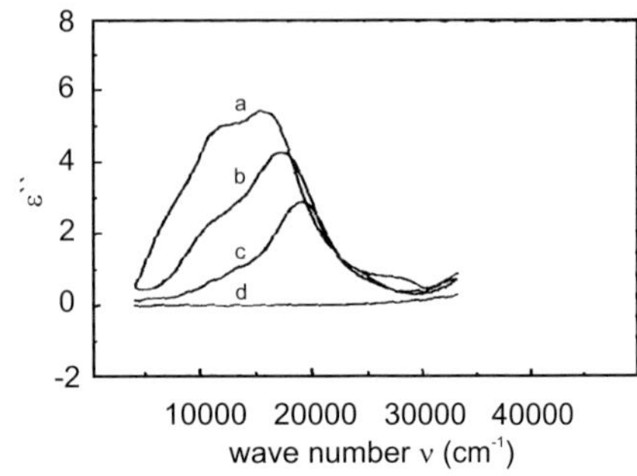

Fig. 6.4. Real part (*left*) and imaginary part (*right*) of the dielectric functions for four plasma polymer multilayers with embedded silver nanoparticles calculated from transmission and reflection spectra. The metal content decreases from curve (a) to curve (c). Curve (d) is the dielectric function of a polymer film without metal nanoparticles

The shift in the spectral position of plasma resonance absorption was also observed for evaporated discontinuous silver films. With increasing particle size ($\bar{D} \geq 5$ nm) and decreasing shape factor, the position shifts to lower wave numbers, but sometimes films are only characterized by an effective thickness determined from the mass deposition [346, 492–496]. It was also observed that the half-width increased [493]. There have only been a few attempts to seek a correlation between the spectral position of the plasma resonance absorption and the particle size or shape of thin silver films [495–499]. In some investigations (e.g., [497]) it becomes obvious that the correlation between particle size and spectral position can be misleading if particle shape is not considered. A constant transmission point was not observed.

For thin insulating films with embedded silver particles, a spectral shift in the plasma resonance absorption was also observed. For example, there is a shift to lower wave numbers with increasing particle size in SiO_2 films [193, 446, 500] or in polyamide films [182] films with silver particles. For KBr films with silver particles, a reflection minimum was found at $\tilde{\nu} = 32\,000$ cm^{-1}. At $\tilde{\nu} \leq 32\,000$ cm^{-1}, reflection increases with particle size [501]. No constant transmission point was found. In [492], transmission spectra of thin silver films with various effective thicknesses are very close together but do not form a constant transmission point. Measurements with polarized light on evaporated silver films [502–504], SiO_2 thin films [446] or gelatin films [505] with embedded silver particles confirm the described spectral dependence of the plasma resonance absorption on particle anisotropies.

For some other insulating materials with embedded silver nanoparticles which were not deposited under vacuum conditions, the spectral position of the plasma resonance absorption to lower wave numbers was found to shift with increasing particle size ($\bar{D} \geq 10$ nm), e.g., gelatin with silver nanoparticles [505]. These films deposited from a colloidal solution are nearly spherical and it is often not necessary to consider the shape factor if particles are not coagulated [506]. With larger filling factors, coagulation takes place. Increasing particle size (e.g., from $\bar{D} \approx 8$ nm to $\bar{D} \approx 65$ nm [505]) causes a substantial red shift. For smaller silver particles ($\bar{D} \leq 10$ nm) embedded in glass, there is no substantial spectral shift but a decrease in half-width with decreasing particle size [507]. More details about the optical properties of colloidal particles fixed in insulating matrix materials are given in [6, 507–509]. Optical properties of cluster material and free metal particles are discussed in [6, 11, 12].

It should be noted that thin discontinuous metal films and embedded metal particles are also the subject of investigations into the nonlinear properties of metal particles. Nonlinear susceptiblities have been determined for discontinuous silver films [510], silver or gold nanoparticles [511–514] and silver particles in glass or in various polymers [515, 516]. Great attention is given to the determination of the electron dynamics using optical pump-probe experiments [517–523].

For plasma polymer films with other embedded metal nanoparticles such as indium, tin, or copper [323, 336], plasma resonance absorptions have low intensity. In particular, for indium and tin nanoparticles, resonances are spread over a large spectral region and show only weak dependence on particle size and shape. Only in the case of embedded copper particles was there a significant spectral shift in the transmission minimum to higher wave numbers with decreasing particle size and increasing shape factor.

6.1.3 Changes in Optical Properties Due to Thermal Treatment

The nanostructural changes described in Chap. 4 lead to substantial changes in the spectral position, half-width and relative intensity of the plasma resonance absorption. Figure 6.5 shows the transmission and reflection spectra of 6 samples of a plasma polymer thin film with embedded silver nanoparticles (film II/5) before and after thermal treatment up to 480 K under vacuum conditions (10^{-3} Pa)

The mean particle size decreases and the mean shape factor increases from spectrum (a) to spectrum (f). The table in Fig. 6.5 gives the spectral position $\tilde{\nu}_r$ [cm^{-1}], the half-width Γ [cm^{-1}] and the relative intensities of the transmission minima I_m. Thermally induced nanostructural changes such as reshaping, coalescence and Ostwald ripening (Sect. 4.1.1) cause the spectral position of the transmission minimum to shift to higher wave numbers. This shift decreases with decreasing mean particle size. In samples (c) and (f), the shifts are about 2 200 cm^{-1} and 1 100 cm^{-1}, respectively. During nanostructural changes, for samples (c) to (f), the half-width Γ of the plasma resonance absorption decreases and the relative intensity I_m increases. In sample (a), which has the largest particles, the plasma resonance absorption found at 11 500 cm^{-1} before thermal treatment is no longer visible after thermal treatment. The s pectrum is comparable to that of a thin embedded silver

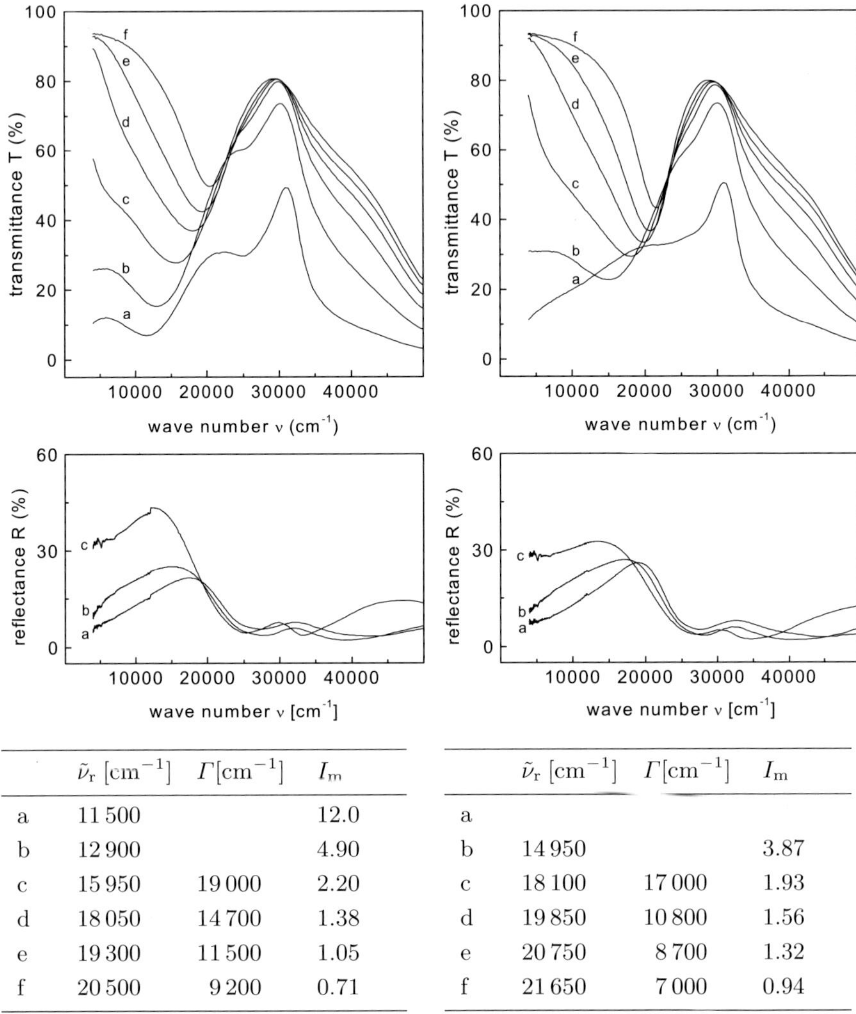

	$\tilde{\nu}_r$ [cm^{-1}]	Γ[cm^{-1}]	I_m		$\tilde{\nu}_r$ [cm^{-1}]	Γ[cm^{-1}]	I_m
a	11 500		12.0	a			
b	12 900		4.90	b	14 950		3.87
c	15 950	19 000	2.20	c	18 100	17 000	1.93
d	18 050	14 700	1.38	d	19 850	10 800	1.56
e	19 300	11 500	1.05	e	20 750	8 700	1.32
f	20 500	9 200	0.71	f	21 650	7 000	0.94

Fig. 6.5. Optical transmission and reflection spectra of a plasma polymer film with embedded silver nanoparticles before (*left*) and after (*right*) thermal treatment up to 480 K. The particle size decreases and the particle shape factor increases from spectrum (a) to (f)

film. The constant transmission point at $22\,000$ cm^{-1} does not shift and is more distinct after thermal treatment.

Because reflection measurements were only possible for larger sample areas, just three spectra are given in Fig. 6.5 for the film before and after thermal treatment. For small particle sizes and large shape factors, reflection depends only weakly on particle size and shape. The reflection spectra does not change substantially during thermal treatment, but at wave numbers

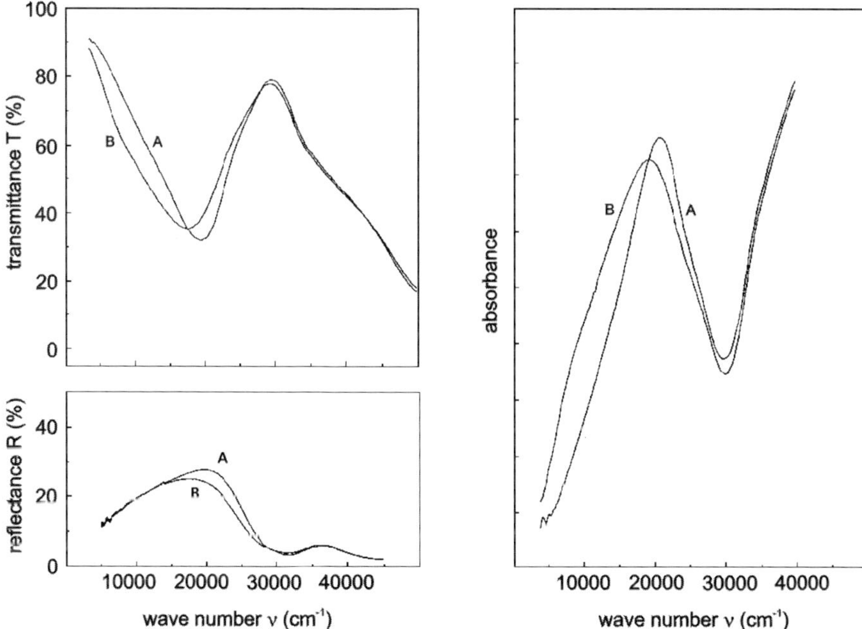

Fig. 6.6. Transmission spectra (*left*), reflection spectra (*bottom left*) and absorption spectra before (B) and after (A) thermal treatment up to 480 K [sample (d) from Fig. 6.5]

lower than $\tilde{\nu} \approx 25\,000$ cm^{-1}, reflection values increase. Similar behavior was found for reflection spectra on silver nanoparticles embedded in glass [525].

For further discussion of spectral changes resulting from nanostructural changes, both spectra from sample (d) were selected from Fig. 6.5 and plotted in one diagram. Figure 6.6 gives the measured transmission and reflection spectra as well as the calculated absorption spectra of sample (d) before and after thermal treatment. The exact congruence of both spectra at wave numbers $\tilde{\nu} \gtrsim 30\,000$ cm^{-1} shows that nanostructural changes are not connected with substantial changes in the polymer matrix. After thermal treatment, film thickness as well as optical properties of the plasma polymer are unchanged. Additionally, it can be assumed that no substantial changes occur in the chemical film structure. Changes in the optical spectra are only due to changes in the particle size and shape distribution. The relative intensity of plasma resonance absorption increases with $\Delta I_\mathrm{m} = 0.18$ and the half-width decreases with $\Delta \Gamma = 3\,900$ cm^{-1}. The spectral shift $\Delta \tilde{\nu}_\mathrm{r} = 200$ cm^{-1} of the transmission minimum to higher wave numbers can be explained by the increase in mean shape factor $\bar{S}$. A final detailed discussion together with

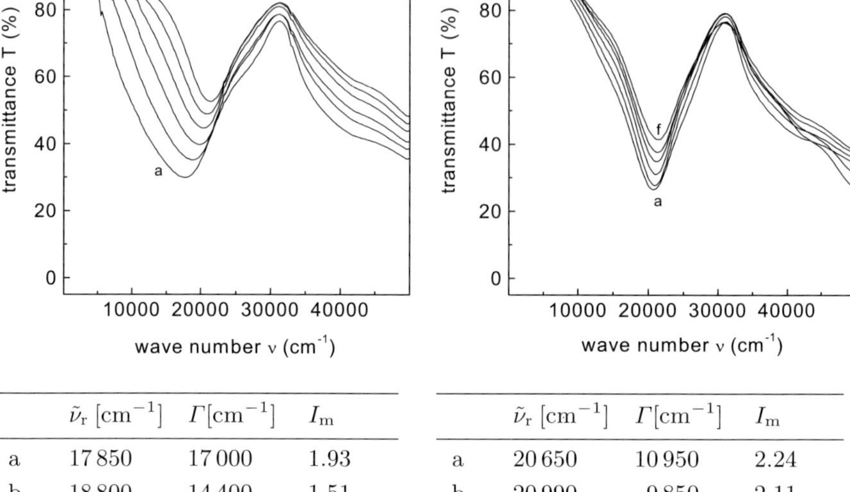

	$\tilde{\nu}_{\mathrm{r}}\,[\mathrm{cm}^{-1}]$	$\Gamma\,[\mathrm{cm}^{-1}]$	I_{m}		$\tilde{\nu}_{\mathrm{r}}\,[\mathrm{cm}^{-1}]$	$\Gamma\,[\mathrm{cm}^{-1}]$	I_{m}
a	17 850	17 000	1.93	a	20 650	10 950	2.24
b	18 800	14 400	1.51	b	20 990	9 850	2.11
c	19 900	12 350	1.20	c	21 200	9 050	1.71
d	20 400	10 550	0.93	d	21 200	9 050	1.40
e	20 900	9 600	0.78	e	21 450	9 050	1.21
f	21 450	8 200	0.64	f	21 450	9 300	0.05

Fig. 6.7. Optical transmission spectra of a plasma polymer film with embedded silver nanoparticles (multilayer system) before (*left*) and after (*right*) laser irradiation at $\lambda = 1\,064$ nm

TEM micrographs will be given after the results of optical calculations at the end of this chapter.

As already mentioned in Sect. 4.3, laser irradiation can induce nanostructural changes. Obviously, these nanostructural changes can alter the spectral position, half-width and relative intensity of the plasma resonance absorption. Figure 6.7 depicts six spectra of a plasma polymer multilayer system with embedded silver nanoparticles (film I/15) before (left) and after (right) laser irradiation at $\lambda = 1\,064$ nm (Nd–YAG laser). The mean particle size decreases and the mean shape factor increases from spectrum (a) to spectrum (f). Before laser irradiation, there is a spectral shift in the plasma resonance absorption from $\tilde{\nu}_{\mathrm{r}} = 17\,850$ cm^{-1} in spectrum (a) to $\tilde{\nu}_{\mathrm{r}} = 21\,450$ cm^{-1} in spectrum (f) with decreasing particle size and increasing particle shape factor. After laser irradiation, the plasma resonance absorption does not change its position so much. The positions $\tilde{\nu}_{\mathrm{r}} = 20\,650$ cm^{-1} in spectrum (a) and $\tilde{\nu}_{\mathrm{r}} = 21\,450$ cm^{-1} in spectrum (f) differ by only 800 cm^{-1}. The half-width

Γ, also given in the table of Fig. 6.7, decreases after laser irradiation and a small increase was found in the relative intensities I_m.

These changes in optical transmission spectra can also occur as a result of thermally induced nanostructural changes. The latter lead to an increase in the mean shape factor $\bar{S}$ and a decrease in the standard deviation σ_D of the mean particle size $\bar{D}$, as was demonstrated in Fig. 4.5. The transmission spectra for wave numbers $\tilde{\nu} \geq 30\,000$ cm^{-1} in Fig. 6.7 demonstrate that the plasma polymer matrix was not destroyed by laser irradiation. Comparing the spectra before and after laser irradiation, there are only slight changes in optical absorption in the spectral region $\tilde{\nu}_\mathrm{r} \geq 40\,000$ cm^{-1}.

The left-hand part of Fig. 6.8 depicts a direct comparison between two spectra of embedded silver nanoparticles before and after laser irradiation (film I/14). There are only slight changes in optical transmission behavior at wave numbers $\tilde{\nu} \geq 30\,000$ cm^{-1}. The plasma resonance absorption shifts from $\tilde{\nu}_\mathrm{r} = 16\,500$ cm^{-1} to $\tilde{\nu}_\mathrm{r} = 21\,000$ cm^{-1} and the half-width decreases substantially from $\Gamma = 21\,500$ cm^{-1} to $\Gamma = 11\,500$ cm^{-1}.

Once again, in the case of plasma polymer multilayer films with embedded gold particles, laser irradiation does not destroy the multilayer system. Figure 6.8 also shows on the right the transmission spectra of two samples before and after laser irradiation (film I/16). There are very small changes in optical transmission at wave numbers $\tilde{\nu} \geq 30\,000$ cm^{-1}. The plasma resonance absorption of the gold nanoparticles found at $\tilde{\nu}_\mathrm{r} = 16\,000$ cm^{-1} does not change position.

As mentioned in Sect. 4.3, optical spectral data confirm the presumption that laser irradiation does not substantially affect the polymer matrix. The multilayer system with embedded silver particles remains intact during laser irradiation. Only the size and shape of the embedded particles were modified by this treatment.

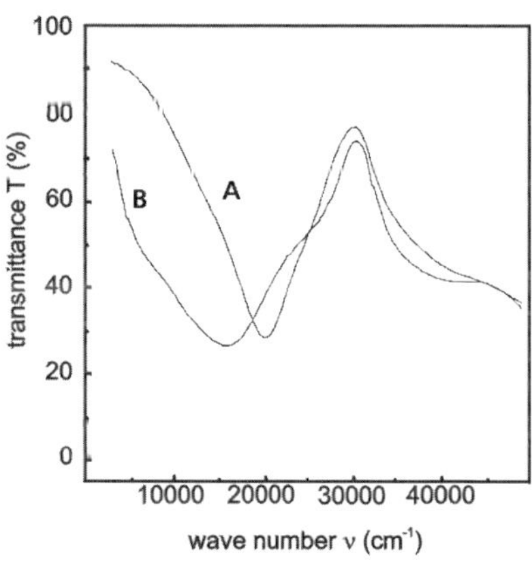

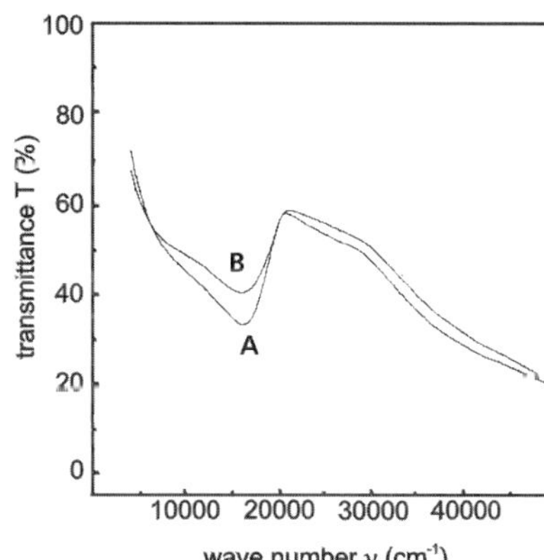

Fig. 6.8. Optical transmission spectra of two plasma polymer films with embedded silver (*left*) and gold (*right*) nanoparticles before (B) and after (A) laser irradiation

A number of of experiments have been performed with the aim of modifing the optical properties of embedded or supported silver particles by laser irradiation, e.g. with a view to creating optical data storage media [526] or to fabricate almost monodisperse metal nanoparticles [527,528] or to change the particle size and shape in limited spot area only [244,529,530].

The nanostructural changes discussed up to now have mainly affected only particle size and shape distributions. The multilayer system consisting of a first plasma polymer film, a plasma polymer film with embedded silver nanoparticles and a second plasma polymer film was preserved. The next task is to discuss optical properties of multilayer systems that have been destroyed during longer thermal treatments at higher temperatures.

Figure 6.9 shows the transmission and reflection spectra for six samples of a plasma polymer multilayer with embedded silver nanoparticles (film II/4) before (left) and after (right) thermal treatment under vacuum conditions (10^{-3} Pa) for 180 min at 570 K. Particle sizes decrease and particle shape factors increase from spectrum (a) to (f).

In contrast to previous results (Figs. 6.5 and 6.7), there is also a substantial change in the optical transmission of the plasma polymer at wave numbers $\tilde{\nu} \geq 30\,000$ cm^{-1}. After thermal treatment, in this spectral region, spectra behave like the spectra of thin, discontinuous silver films deposited without polymer matrix [531]. It must be noted that, because of thermal desorption, the plasma polymer film is no longer present, or only a very small part of it remains. This agrees with the XPS investigation in Fig. 4.8.

As observed before the film from Fig. 6.9, the plasma resonance absorption shifts to higher wave numbers and the half-width decreases. After thermal treatment, the constant transmission point is found at $22\,500$ cm^{-1}. To visualize the changes in spectral position $\tilde{\nu}_{\mathrm{r}}$ [cm^{-1}] and half-width Γ [cm^{-1}], Fig. 6.10 shows data taken from twelve spectra plotted as a function of the relative sample position (see the deposition reactor in Fig. 2.10). With increasing position relative to the substrate, mean particle sizes increase and shape factors decrease. Points marked with letters correspond to spectra from Fig. 6.9.

The spectral shift $\Delta\tilde{\nu} = 1\,000\text{--}3\,000$ cm^{-1} of the plasma resonance absorption corresponds to the values in Figs. 6.5 and 6.7. The decrease in half-width Γ is much greater than previously observed. The half-width decreases down to half the value before thermal treatment. Half-width values of $\Gamma \leq 4\,000$ cm^{-1} were not observed in any untreated polymer films with embedded silver nanoparticles. Similar low values ($\Gamma \approx 3\,000$ cm^{-1}) were only found for free silver clusters [532] or silver colloids in KBr [533].

Thermal desorption of the polymer matrix meant that it was not possible to determine the nanostructure of the samples from Fig. 6.9 using TEM micrographs. The particle size and shape distributions could not therefore be ascertained. On the basis of nanostructural studies of other films, it is assumed that silver particles are nearly spherical with a narrow size and

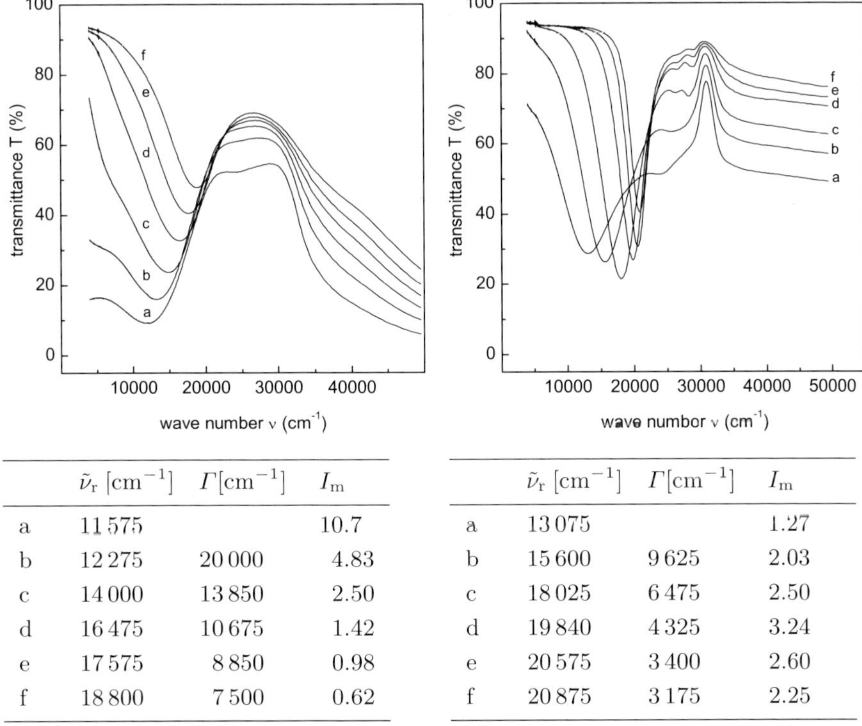

	$\tilde{\nu}_r\,[\mathrm{cm}^{-1}]$	$\Gamma\,[\mathrm{cm}^{-1}]$	I_m
a	11 575		10.7
b	12 275	20 000	4.83
c	14 000	13 850	2.50
d	16 475	10 675	1.42
e	17 575	8 850	0.98
f	18 800	7 500	0.62

	$\tilde{\nu}_r\,[\mathrm{cm}^{-1}]$	$\Gamma\,[\mathrm{cm}^{-1}]$	I_m
a	13 075		1.27
b	15 600	9 625	2.03
c	18 025	6 475	2.50
d	19 840	4 325	3.24
e	20 575	3 400	2.60
f	20 875	3 175	2.25

Fig. 6.9. Optical transmission spectra of a plasma polymer thin film with embedded silver particles before (*left*) and after (*right*) thermal treatment (180 min at 570 K)

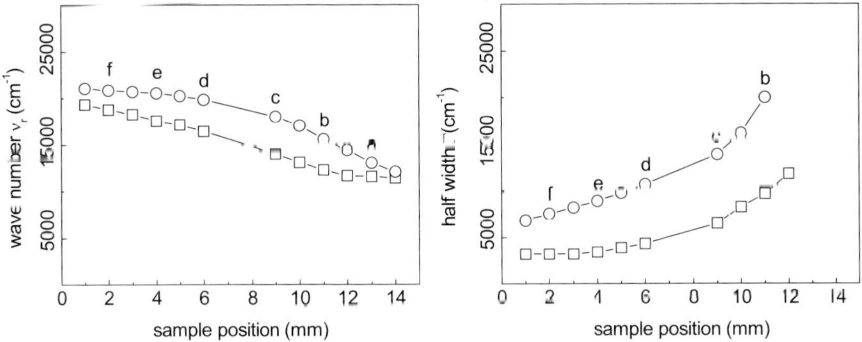

Fig. 6.10. Spectral position of plasma resonance absorption $\tilde{\nu}_r\,[\mathrm{cm}^{-1}]$ (*left*) and half width $\Gamma\,[\mathrm{cm}^{-1}]$ (*right*) as a function of relative sample position in the deposition reactor before (□) and after (○) thermal treatment (180 min at 570 K). Letters correspond to spectra from Fig. 6.9

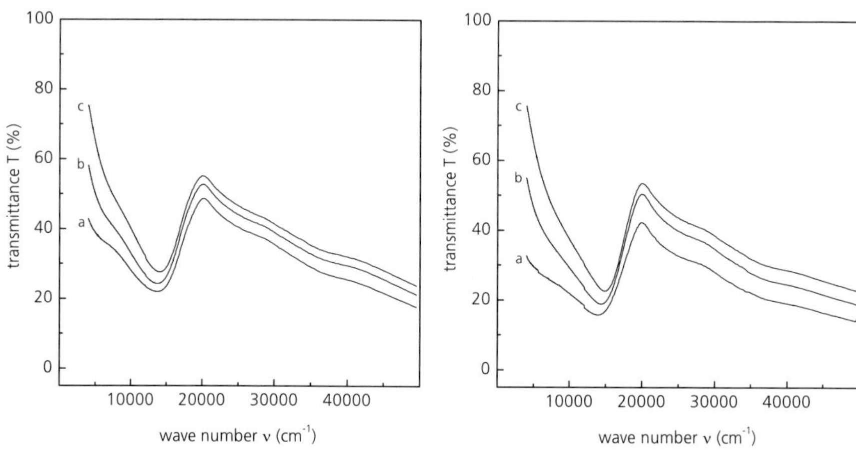

Fig. 6.11. Optical transmission spectra of a plasma polymer thin film with embedded gold particles before (*left*) and after (*right*) thermal treatment (15 min at 480 K)

shape distribution. Furthermore, the occurrence of a constant transmission point suggests the formation of a concentric polymer shell around the silver particles (see Sect. 6.2.4). That there is some polymer left was demonstrated by the XPS investigations in Fig. 4.8.

All the demonstrated changes in optical spectra due to thermal treatment can be explained in terms of diffusion processes of silver atoms around and between silver particles. In particular, coalescence and reshaping lead to increased particle sizes and increased shape factors (Fig. 4.6). The result is a shift in the plasma resonance absorption to higher wave numbers (blue shift) and a decrease in half-width. For as-deposited films with increasing particle size, the plasma resonance absorption shifts to lower wave numbers (red shift). This contradiction can only be resolved if the particle shape factor $\bar{S}$ is considered as the main parameter responsible for the spectral shift. As mentioned later, in Sect. 6.2.4 and Fig. 6.19, the shape factor S also increases the depolarization factor $0 < L \leq 1/3$. The resonance frequency $\tilde{\nu}_{\mathrm{r}}$ given in (6.7) shifts to higher wave numbers.

The idea that the increased particle shape factor is responsible for the spectral shift due to thermal treatment agrees well with other experimental data. The strong dependence of the spectral position of plasma resonance absorption on particle shape was also observed for deformed silver or gold particles embedded in glass [534, 535]. Furthermore, experimental data for silver particles embedded in polymer films will be confirmed by the optical calculation in the following sections.

If a constant transmission point was measured for films after deposition, it was also found after thermal treatment. There is only a slight shift in the spectral position of the constant transmission point to higher wave numbers.

The constant transmission point cannot be compared with the so-called X-point $\tilde{\nu} = 35\,000$ cm^{-1} found in the optical absorption spectra of bulk silver. The latter results from the temperature dependence of the dielectric function for silver [536]. In transmission spectra measured during the photo-doping of chalcogenide films $(Ge_{30}S_{70})$ with silver, an isotransmission point was found at $\tilde{\nu} = 20\,800$ cm^{-1} [537]. There is a photo-induced diffusion of a silver film inside the chalcogenide film. The nanostructure of this material has not been observed.

Having discussed changes in optical transmission spectra for plasma polymer films with embedded silver particles, let us now consider changes in the optical properties of polymer films with embedded gold particles. Figure 6.11 shows the optical spectra of a plasma polymer multilayer system with embedded gold nanoparticles (film I/17) before (left) and after (right) thermal treatment (15 min at 480 K). The particle size decreases and the particle shape factor increases continuously from spectrum (a) to spectrum (c).

A spectral shift of the plasma resonance absorption to lower wave numbers is observed for embedded gold particles as particle size increases and shape factor decreases, provided the particles are small enough. In a like manner to thin silver films, thin discontinuous gold films also show a spectral shift in the plasma resonance absorption with increasing effective thickness, a change comparable with increasing particle size and decreasing shape factor [347, 538]. After thermal treatment, there is little or no spectral shift in the plasma resonance absorption to higher wave numbers. The spectral shift is much smaller than for silver thin films. A red shift was found in the case of plasma polymer films with embedded gold nanoparticles deposited by simultaneous plasma polymerization and metal sputtering or by co-sputtering [258, 277, 299].

Due to nanostructural changes during thermal treatment, substantial changes occur in optical spectra of embedded silver nanoparticles. Silver particles show the strongest dependence of optical properties on particle size and shape distributions. The shift in the spectral position of the plasma resonance absorption can only be described if changes in particle shape are considered.

6.2 Optical Calculations

6.2.1 Modelling Approaches and Dielectric Functions

Modelling Approaches. The starting point for all optical calculations relating to embedded nanoparticles is to select the dielectric functions of the metal $\hat{\varepsilon}_{me}(\tilde{\nu})$ and the insulating polymer material $\hat{\varepsilon}_{po}(\tilde{\nu})$. For the metal, dielectric functions for the bulk material or experimentally determined dielectric functions for thin films can be used. For polymer films, it is also possible to use a constant dielectric function or experimental data.

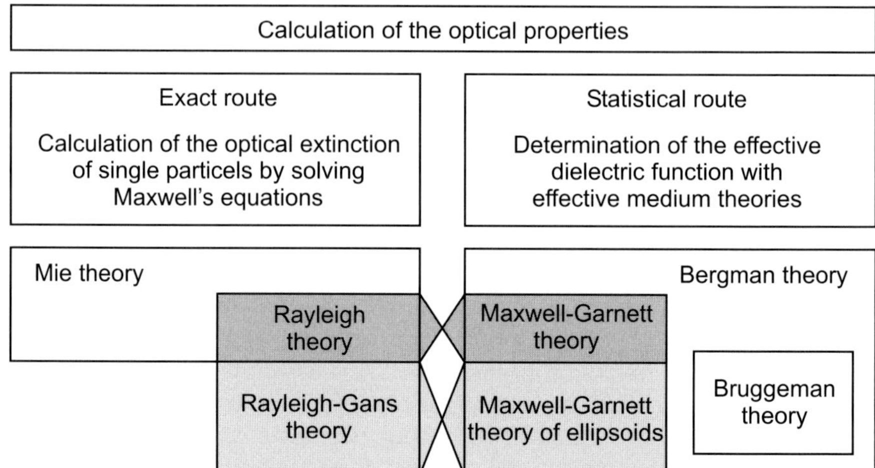

Fig. 6.12. Routes available for calculating optical properties of insulator thin films with embedded nanoparticles

The success of the calculations depends mainly on how nanostructural information is introduced. In both approaches to modelling, nanostructural information can be introduced at different levels. On the first level, only the metal particle filling factor or a mean particle size is introduced. Then the particle shape may also be introduced. The best representation of the nanostructure can be achieved by introducing the particle size and shape distribution or the size and shape of each individual particle into the optical theory.

In order to calculate optical properties of polymer films with embedded particles, two routes can be selected (see Fig. 6.12). In the exact route, the extinction cross sections $C_{\mathrm{ext}}(\tilde{\nu})$ of single particles are calculated by solving Maxwell's equations for spherical particles (Mie theory). To reduce the effort involved in these calculations, the quasistatic approximation (Rayleigh theory) can often be used. This considers only dipole excitations. The Rayleigh–Gans theory is a quasistatic approximation for non-spherical particles and includes the Rayleigh theory as a special case. The calculated extinction spectra for single particles, or better, a summation of various excitation spectra for a particle assembly, can be compared with the experimental spectra of embedded nanoparticles.

In the statistical route, an effective dielectric function $\hat{\varepsilon}(\tilde{\nu})$ is calculated from the dielectric functions of the metal $\hat{\varepsilon}_{\mathrm{me}}(\tilde{\nu})$ and the polymer material $\hat{\varepsilon}_{\mathrm{po}}(\tilde{\nu})$ using a formula specified by some effective medium theory. The most general effective medium theory is the Bergman theory in which the nanostructure of the composite material is treated using a spectral density function. Bergman theory includes the solutions from the Bruggeman theory and the

Maxwell–Garnett theory for spherical, parallel oriented and randomly oriented ellipsoidal particles.

The effective dielectric function can be compared with dielectric functions calculated from experimental transmission and reflection spectra. Better still, optical transmission, absorption or reflection spectra can be calculated exactly from the dielectric function $\hat{\varepsilon}(\tilde{\nu})$ using the Fresnel formulas [5] for the optical system air–composite media–quartz substrate. The thickness of the film with embedded particles and the thickness of other layers present which do not contain metal nanoparticles have to be included as further parameters. Calculated transmission and reflection spectra can be compared directly with experimental spectra.

Both routes give comparable results. The only difference between Rayleigh theory and Maxwell–Garnett theory for spherical particles is the different mathematical formulation. The calculated spectra are identical. In principle, this is also the case when Rayleigh–Gans theory and Maxwell–Garnett theory are used to treat parallel oriented elliposidal particles. The difference lies only in the way nanostructural information is introduced. In the Rayleigh–Gans theory the size and shape of each individual particle is given, whilst in the Maxwell–Garnett theory, the statistical particle size and shape distributions for the whole particle assembly are introduced. Nevertheless, different aspects of the optical properties of polymer films with embedded nanoparticles can be discussed using the two routes.

Dielectric Functions. An adequate choice of dielectric function for the basic materials is one of the main prerequisites for successfully calculating optical properties. As already mentioned in Sect. 6.1.1, the dielectric function for the metals can be calculated using the Drude–Sommerfeld theory (6.3). In principle, the intensity of the calculated plasma resonance absorption is too high [305]. This is also the case if the reduced, particle-size-dependent collision frequency

$$\tilde{\nu}_{\mathrm{c}}^{*}(R) = \tilde{\nu}_{\mathrm{c}} + \frac{v_{\mathrm{F}}}{cR} \tag{6.9}$$

is used [539], where v_{F} is the Fermi velocity, c the velocity of light and R the particle radius.

Otherwise, experimentally determined dielectric functions $\hat{\varepsilon}_{\mathrm{me}}(\tilde{\nu})$ can be used [540]. The values due to Johnson and Christy [541], which are used very frequently, were determined from transmission and reflection measurements on thin metal films in a spectral region between $5\,100 \leq \tilde{\nu} \leq 53\,000$ cm^{-1}. Dielectric functions determined from measurements on silver particle systems with different mean particle sizes have also been calculated for wave numbers $17\,500 \leq \tilde{\nu} \leq 36\,000$ cm^{-1} [485]. Experimental values for a large spectral range (1–600 eV) are given by synchrotron measurements [542].

The left-hand part of Fig. 6.13 gives the dielectric function for silver $\hat{\varepsilon}_{\mathrm{me}}(\tilde{\nu})$ after Johnson and Christy [541]. This dielectric function is used to cal-

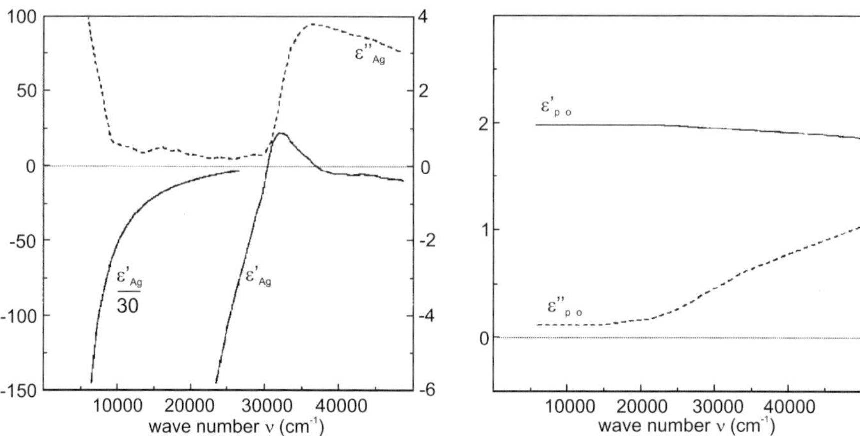

Fig. 6.13. *Left*: dielectric function of silver $\hat{\varepsilon}_{\mathrm{Ag}}(\tilde{\nu})$ after Johnson and Christy. *Right*: dielectric function of the plasma polymer $\hat{\varepsilon}_{\mathrm{po}}(\tilde{\nu})$ determined from experimental data (Fig. 2.7)

culate optical properties of plasma polymer thin films with embedded metal nanoparticles in Sect. 6.2.4.

For plasma polymer thin films, various values were calculated for the dielectric functions of films made from benzene under different deposition conditions (see Fig. 2.7). To simplify matters and because calculations are mainly influenced by the dielectric function of the metal, a dielectric function $\hat{\varepsilon}_{\mathrm{po}}(\tilde{\nu})$ with constant real part $\varepsilon_{\mathrm{me}}(\tilde{\nu})' = 2.8$ and imaginary part $\varepsilon_{\mathrm{me}}(\tilde{\nu})''$ that increases with increasing wave number is used for the calculation in Sect. 6.2.4 (Fig. 6.13).

6.2.2 Calculations in the Exact Approach

Rayleigh–Gans Theory. As rigorous solutions for extinction and scattering are available for only a few particle geometries, e.g., spheres (Mie theory) [484], spheroids [486] or infinite cylinders [487,488], spectra for particles with other shapes must be calculated by approximations. In the so-called Rayleigh approximation, in which particles are small compared to the wavelength of the incident light, Gans [543] developed a model for ellipsoidal particles and Fuchs [544] another for cubic particles. Both begin by calculating the dipole polarizability of the particle.

In the Rayleigh approximation the particle is so small that an applied electric field $\boldsymbol{E}_0$ only induces a dipole with dipole moment $\boldsymbol{p}$ in the particle. In general, $\boldsymbol{E}_0$ and $\boldsymbol{p}$ are not collinear, meaning that the polarizability $\hat{\alpha}$ of the particle is a tensor:

$$\boldsymbol{p} = \varepsilon_0 \varepsilon_{\mathrm{po}} \hat{\alpha} \boldsymbol{E}_0 \ . \tag{6.10}$$

In this notation, the elements of the polarizability tensor are only proportional to the particle volume, as happens for the atomic polarizability. $\varepsilon_{\mathrm{po}}$ is the dielectric constant of the surrounding polymer medium. A dipole with linear response oscillates with the same frequency as the applied electric field and hence emits an electric field $\boldsymbol{E}_{\mathrm{s}}$ given asymptotically in the far field by

$$\boldsymbol{E}_{\mathrm{s}} = \frac{e^{ikr}}{ikr} \frac{ik^3}{4\pi\varepsilon_0\varepsilon_{\mathrm{po}}} \boldsymbol{e}_r \times (\boldsymbol{e}_r \times \boldsymbol{p}) \ . \tag{6.11}$$

Using Poynting's theorem, the extinction cross section C_{ext} for the dipole is obtained as

$$C_{\mathrm{ext}} = k \ \mathrm{Im} \sum_{i=1}^{3} \sum_{j=1}^{3} \alpha_{ij} e_i e_j \ , \tag{6.12}$$

where Im means the imaginary part, and e_i and e_j are the normalized components of the incident field vector. For an ellipsoidal particle with three different elementary axes A, B and C, the polarizability tensor only has diagonal elements α_{jj}, with

$$\alpha_{jj} = 4\pi ABC \frac{\varepsilon_{\mathrm{me}}(\tilde{\nu}) - \varepsilon_{\mathrm{po}}(\tilde{\nu})}{\varepsilon_{\mathrm{po}}(\tilde{\nu}) + G_j[\varepsilon_{\mathrm{me}}(\tilde{\nu}) - \varepsilon_{\mathrm{po}}(\tilde{\nu})]} \ , \tag{6.13}$$

where G_j are factors taking into account the geometry of the ellipsoid. The latter satisfy the conditions

$$\sum G_j = 1 \ , \quad G_j \le 1 \quad \forall j \ . \tag{6.14}$$

For comparison with experiment, the special case of spheroids has been found suitable. They are characterized by coincidence of two of the three elementary axes. One distinguishes elongated or prolate spheroids with $B = C$, and flattened or oblate spheroids with $A = B$. In the following, A is always the major axis and B the minor axis of the spheroid. Accordingly, only two factors $G_1(e)$ and $G_2(e)$ are relevant, depending on the eccentricity e, defined for a spheroid as

$$e^2 = 1 - \left(\frac{B}{A}\right)^2 \ . \tag{6.15}$$

G_1 differs for prolate and oblate spheroids:

$$G_1(e) = -g(e)^2 \left(1 - \frac{1}{2e}\ln\frac{1+e}{1-e}\right) \quad \text{(prolate)} \ , \tag{6.16}$$

$$G_1(e) = \frac{g(e)}{2e^2}\left[\frac{\pi}{2} - \tan(g(e))^{-1}\right] - \frac{g(e)^2}{2} \quad \text{(oblate)} \ , \tag{6.17}$$

where

$$g^2(e) = \frac{1 - e^2}{e^2} \ . \tag{6.18}$$

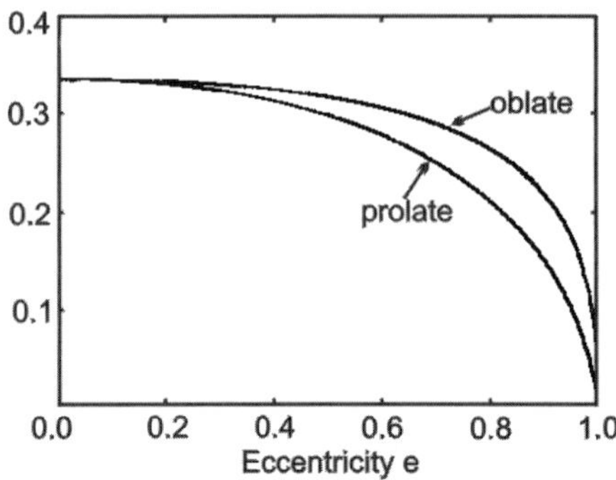

Fig. 6.14. Geometric factor $G_1(e)$ as a function of eccentricity e

In Fig. 6.14, $G_1(e)$ is plotted against eccentricity e for both prolate and oblate spheroids. It starts with $G_1(0) = 1/3$, which is the case of a spherical particle ($A = B = R$) and decreases with increasing eccentricity. On the other hand, following from (6.14), $G_2(e)$ also starts with $G_2(0) = 1/3$, but it increases with increasing eccentricity. It follows now from (6.13) that, for a spheroidal silver particle, two surface plasma resonances are induced, namely, when the optical constants of silver satisfy the conditions

$$\varepsilon(\tilde{\nu}) = -\varepsilon_{\mathrm{po}} \frac{1 - G_j}{G_j} \ . \tag{6.19}$$

To simplify matters, consider the dielectric constant to be a purely Drude dielectric constant without damping, i.e.,

$$\varepsilon(\tilde{\nu}) = 1 - \frac{\tilde{\nu}_{\mathrm{p}}^2}{\tilde{\nu}^2} \ , \tag{6.20}$$

where $\tilde{\nu}_{\mathrm{p}}$ is the plasma frequency of the free electrons. It then becomes obvious that the plasma resonance for G_1 contributes to the spectrum at lower frequencies and the plasma resonance for G_2 contributes at higher frequencies than the peak frequency of the plasma resonance of the equivalent volume sphere with $G_1 = G_2 = 1/3$.

Results of Calculations in the Exact Approach. The calculation will be demonstrated for two different particle assemblies found on the TEM micrograph of Fig. 4.26. The TEM micrograph shows silver particles which are representative of the as-deposited film and the thermally treated film, respectively. The particles were assumed to be rotational ellipsoids with major axis A and minor axis B (see histograms in Fig. 4.27).

The optical extinction spectra for prolate spheroids were computed using optical constants of silver from [545, 546], taking into account additional damping of the plasma resonance absorption according to the mean free path model. The left-hand part of Fig. 6.15 gives a section enlargement of Fig. 4.26 from the as-deposited nanostructure. Five particles are marked. The major

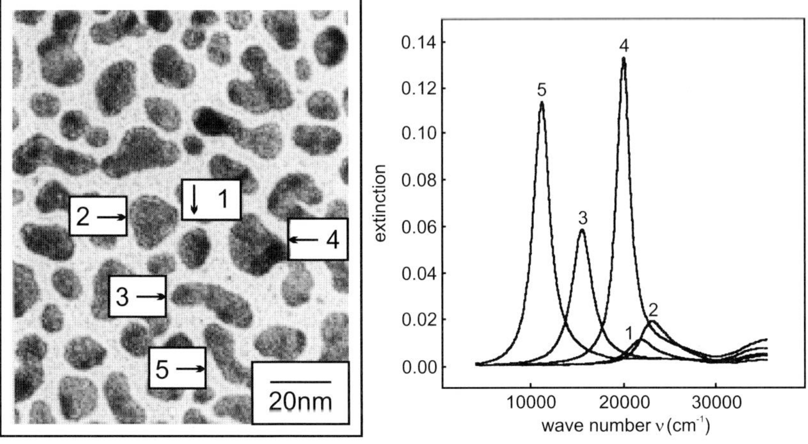

Fig. 6.15. Extinction spectra of single particles

axis A, the minor axis B and eccentricities e are given as follows:

$$\begin{array}{llll}
\text{Particle 1} & A = 9.7 \text{ nm} & B = 7.3 \text{ nm} & e = 0.66 \\
\text{Particle 2} & A = 12.1 \text{ nm} & B = 11.0 \text{ nm} & e = 0.41 \\
\text{Particle 3} & A = 21.0 \text{ nm} & B = 7.0 \text{ nm} & e = 0.94 \\
\text{Particle 4} & A = 23.6 \text{ nm} & B = 13.1 \text{ nm} & e = 0.83 \\
\text{Particle 5} & A = 33.9 \text{ nm} & B = 6.8 \text{ nm} & e = 0.98 \\
\end{array}$$

The right-hand part of Fig. 6.15 shows the calculated extinction spectra of these five selected particles. Peak position and magnitude of the G_1 mode obviously depend on the shape and size of the corresponding particle. The larger the eccentricity of the particle, the lower the wave number where the plasma resonance absorption of the G_1 mode peaks. The peak magnitude correlates approximately with the particle volume, but for a quantitative comparison the dispersion of the optical constants in the polarizability must be taken into account. For this reason, the larger particle 5 has a smaller absorption peak magnitude than particle 4, although it has a larger volume than particle 4.

In the left-hand part of Fig. 6.16, computed extinction spectra are presented for the particle assemblies before and after nanostructural changes. The results from image analysis are used to compute extinction spectra for 368 particles with varying size parameters, adding them up to total spectra of the sample before (solid line) and after thermal treatment (dashed line). The spectra are normalized for better comparison. The right-hand part of Fig. 6.16 gives experimental extinction spectra (film II/7) calculated from transmission spectra for the sample before and after thermal treatment which has a comparable nanostructure to the one depicted in Fig. 4.26.

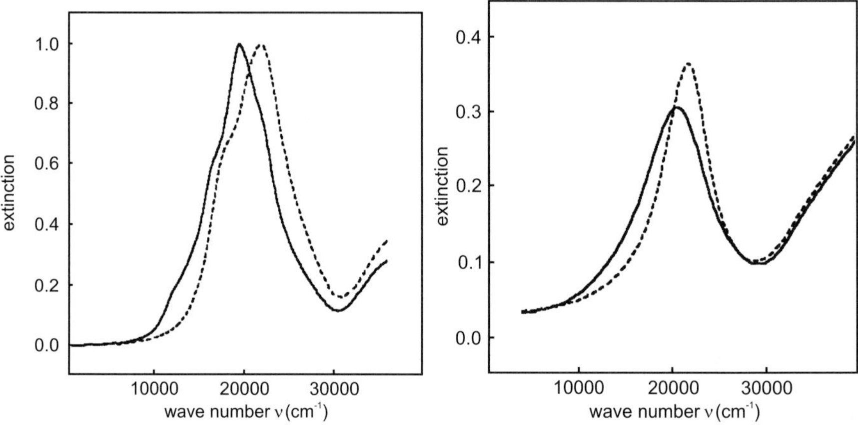

Fig. 6.16. *Left*: calculated extinction spectra added up for 368 particles before (*solid line*) and after (*dashed line*) thermal treatment. *Right*: experimental extinction spectra before (*solid line*) and after (*dashed line*) thermal treatment

It is obvious that the calculated spectrum of the sample after thermal treatment is blue-shifted with a shift of $\Delta\tilde{\nu}_r = 1\,570$ cm^{-1}. The peak positions before and after thermal treatment are $\tilde{\nu}_r = 19\,420$ cm^{-1} and $\tilde{\nu}_r = 21\,050$ cm^{-1}, in good agreement with our experimental data. In the experimental spectra, the plasma resonance absorption peaks at $\tilde{\nu}_r = 20\,410$ cm^{-1} before and $\tilde{\nu}_r = 21\,740$ cm^{-1} after thermal treatment. This means that the experimental extinction peak shifts by $\Delta\tilde{\nu}_r = 1\,330$ cm^{-1} to higher wave numbers. The half-widths with $\Delta\tilde{\nu}_r = 7\,200$ cm^{-1} and $\Delta\tilde{\nu}_r = 6\,730$ cm^{-1} are larger than the measured half-widths $\Gamma = 5\,450$ cm^{-1} and $\Gamma = 4\,130$ cm^{-1}, respectively. One possible reason for this divergence is that the particles which have a more complicated shape are projected to ellipsoids. Another reason may be that the computed spectra exhibit resonances at lower wave numbers which are missing in the measured spectra of Fig. 6.16. They are caused by the limited validity of the model used in our computations. For silver particles with sizes exceeding about 20 nm, the Rayleigh approximation is no longer valid and more accurate computations are needed.

The agreement between experimental and calculated spectra is rather good. The calculations give a good explanation for the blue shift of the plasma resonance absorption of silver nanoparticles in terms of their nanostructural changes during thermal treatment. In computations with the Rayleigh–Gans theory, the values for major and minor particle half-axes were used for each individual particle. Nanostructural information from the TEM micrographs was very well introduced into the optical calculations.

6.2.3 Calculations with Effective Medium Theories

Effective Medium Theories. Effective medium theories can be applied to a large number of nanocomposite materials. In addition to the volume filling factor f , some theories include parameters which describe nanostructure, e.g., the depolarization factor L or the spectral density $g(x)$. If the nanostructure is unknown, these parameters and the filling factor act as fit parameters. If the calculations fit well to experimental data, it is possible to get information about unknown nanostructure, although there are situations where this can also be misleading.

The effective medium theories described in Sect. 6.2.3 have other uses apart from the calculation of optical properties in nanocomposite materials. They can also be used to calculate effective dielectric functions in the microwave region or the effective dielectric constant [547, 548]. In particular, the Hashin–Shtrikman bounds and the Bruggeman theory can be used as mixing formulas for the d.c. conductivity [432, 547, 549, 550], or to determine the effective thermal conductivity and effective thermal expansion coefficients [416]. The Hashin–Shtrikman bounds were also used to determine effective mechanical properties of composite media [550, 551].

The following discussion of effective medium theories requires some further assumptions. The two basic materials are still assumed to be homogeneous. Intermediate layers between them are neglected and can be introduced only by assuming a third material. The latter is quite feasible in some effective medium theories [552].

The most popular effective medium theories are the Maxwell–Garnett theories, derived from classical scattering theory and the Bruggeman theory. With these theories, an effective dielectric function is calculated from the dielectric functions of both basic materials using the volume filling factor. In some extensions of these theories, a unique particle shape is assumed for all particles. There is also another concept based on bounds for the effective dielectric functions. The bounds are valid for a specific nanostructure. Between these bounds, the effective dielectric function varies with the nanostructure of the material. The Bergman theory includes a spectral density function $g(x)$ which is used as a fit function and correlates with the nanostructure of the material.

Maxwell–Garnett theory:

$$\frac{\hat\varepsilon(\tilde\nu) - \hat\varepsilon_{\mathrm{po}}(\tilde\nu)}{\hat\varepsilon(\tilde\nu) + 2\hat\varepsilon_{\mathrm{po}}(\tilde\nu)} = f\,\frac{\hat\varepsilon_{\mathrm{me}}(\tilde\nu) - \hat\varepsilon_{\mathrm{po}}(\tilde\nu)}{\hat\varepsilon_{\mathrm{me}}(\tilde\nu) + 2\hat\varepsilon_{\mathrm{po}}(\tilde\nu)} , \tag{6.21}$$

where $\hat\varepsilon_{\mathrm{me}}(\tilde\nu)$ and $\hat\varepsilon_{\mathrm{po}}(\tilde\nu)$ are the dielectric functions of the metal and the insulator, respectively, $\hat\varepsilon(\tilde\nu)$ is the effective dielectric function, and f is the volume filling factor.

In 1906, Garnett [553] devised the Maxwell–Garnett equation (6.21) to describe the color of metal colloid glasses and thin metal films. Equation (6.21) can be derived from the theory of Rayleigh scattering for spherical particles [554], or from the Lorentz–Lorenz assumption for the electric field of a sphere and the Clausius–Mossotti equation (6.42), using the polarizability of a metal particle when only dipole polarization is considered (6.4). Equation (6.21) was developed for spherical particles and neglects interactions with neighboring particles. All particles are completely encapsulated in an insulating matrix material. It was further assumed that particle sizes are small in comparison with the wavelength of incident light. At higher filling factors, particle–particle interactions can no longer be neglected. This limits the exact validity of (6.21) to small filling factors, e.g., $f < 0.3$ [555].

For thin, discontinuous metal films or polymer thin films with embedded metal nanoparticles, particles do not generally have a spherical shape. Various extensions of the Maxwell–Garnett theory to thin metal–insulator composites with non-spherical particles have been introduced. In most cases, the particles are spheroidal with the same shape (ratio of half-axes a and b) but with different sizes still in the wavelength limit. It remains only to choose between a common orientation or a random orientation for the major axes of the ellipsoids.

In extensions of the Maxwell–Garnett theory to parallel oriented ellipsoids (6.22) [193, 556], all particles are ellipsoids with parallel major axis. Only one depolarization factor L is needed. L describes the ratio between the half-axes of the ellipsoids, and values for L with $0 \leq L \leq 1/3$ (rod), $L = 1/3$ (sphere) and $1/3 \leq L \leq 1$ (disc) are possible. For the extreme geometries $L = 0$ (parallel rods) and $L = 1$ (parallel discs), equation (6.22) is equal to the Wiener bounds (6.27) and (6.28), respectively. For encapsulated spherical particles ($L = 1/3$), (6.22) gives the same result as the Maxwell–Garnett theory, viz., (6.21). For thin films, the ellipsoid orientation relative to the substrate is parallel or perpendicular to the plane of the incident light. A diagonal orientation of the ellipsoidal particles relative to the substrate cannot be considered.

Maxwell–Garnett theory for parallel oriented ellipsoids:

$$\frac{\hat{\varepsilon}(\tilde{\nu}) - \hat{\varepsilon}_{\mathrm{po}}(\tilde{\nu})}{\hat{\varepsilon}_{\mathrm{po}}(\tilde{\nu}) + L\left[\hat{\varepsilon}(\tilde{\nu}) - \hat{\varepsilon}_{\mathrm{po}}(\tilde{\nu})\right]} = f \frac{\hat{\varepsilon}_{\mathrm{me}}(\tilde{\nu}) - \hat{\varepsilon}_{\mathrm{po}}(\tilde{\nu})}{\hat{\varepsilon}_{\mathrm{po}}(\tilde{\nu}) + L\left[\hat{\varepsilon}_{\mathrm{me}}(\tilde{\nu}) - \hat{\varepsilon}_{\mathrm{po}}(\tilde{\nu})\right]} , \quad (6.22)$$

where L is the depolarization factor.

Extensions of the Maxwell–Garnett theory to randomly oriented ellipsoids (6.23) [557] require three depolarization factors L_1, L_2, L_3 with $\sum L_i = 1$ to describe the embedded ellipsoids. It is assumed that the ellipsoids have rotational symmetry so that $L_2 = L_3$. Extreme geometries are rods with $L_1 \gg L_2 = L_3$ and discs with $L_1 \ll L_2 = L_3$. For $L_i = 1/3$, extensions of the Maxwell–Garnett theory to randomly oriented ellipsoids as described by (6.23) coincide with the Maxwell–Garnett equation (6.21). For $L_1 = 1$, equation (6.23) is equal to the second Hashin–Shtrikman bound (6.30).

There is another extension of the Maxwell–Garnett theory to randomly oriented ellipsoids [482] which differs from (6.23). Other extensions of the Maxwell–Garnett theory exist for chiral aggregates of spheres [558], thin films with columnar structures [559], spherical chiral inclusions [558], and embedded spherical particles made from several metals [552].

Maxwell–Garnett theory extended to randomly oriented ellipsoids:

$$(1 - f)\frac{\hat{\varepsilon}(\tilde{\nu}) - \hat{\varepsilon}_{\mathrm{po}}(\tilde{\nu})}{\hat{\varepsilon}_{\mathrm{me}}(\tilde{\nu}) - \hat{\varepsilon}(\tilde{\nu})} = \frac{f}{3}\sum_{i=1}^{3}\frac{\hat{\varepsilon}_{\mathrm{po}}(\tilde{\nu})}{\hat{\varepsilon}_{\mathrm{po}}(\tilde{\nu}) + L_i\left[\hat{\varepsilon}_{\mathrm{me}}(\tilde{\nu}) - \hat{\varepsilon}_{\mathrm{po}}(\tilde{\nu})\right]}\,,$$

$$(6.23)$$

where L_1, L_2, L_3 are depolarization factors with $\sum_{i-1}^{3} L_i = 1$.

In contrast to the Maxwell–Garnett theory, the Bruggeman theory [560] treats both materials on an equal basis. The two basic materials consist of spherical particles without encapsulation of one material by the other. The theory is thus symmetrical. This means that $\hat{\varepsilon}_{\mathrm{me}}(\tilde{\nu})$ and $\hat{\varepsilon}_{\mathrm{po}}(\tilde{\nu})$ can be exchanged in (6.24). Unlike the Maxwell–Garnett theory, the Bruggeman theory gives a percolation threshold at $f_{\mathrm{p}} = 1/3$.

Bruggeman theory:

$$\frac{\hat{\varepsilon}_{\mathrm{me}}(\tilde{\nu}) - \hat{\varepsilon}(\tilde{\nu})}{\hat{\varepsilon}_{\mathrm{me}}(\tilde{\nu}) + 2\hat{\varepsilon}(\tilde{\nu})} = (f - 1)\frac{\hat{\varepsilon}_{\mathrm{po}}(\tilde{\nu}) - \hat{\varepsilon}(\tilde{\nu})}{\hat{\varepsilon}_{\mathrm{po}}(\tilde{\nu}) + 2\hat{\varepsilon}(\tilde{\nu})}\,. \qquad (6.24)$$

Bruggeman developed another effective medium theory for particles with lamellar shape and an asymmetric effective medium theory for spherical particles [433, 560]. Besides the filling factor, neither of these effective medium theories includes any further parameter to represent particle size.

Extensions to parallel oriented or randomly embedded ellipsoidal metal particles can be made in the same way as in the Maxwell–Garnett theory. In the extension for parallel oriented ellipsoids (6.25), the depolarization factor L is used again. The percolation threshold f_{p} shifts to $f_{\mathrm{p}}(L) = L$.

Bruggeman theory extended to parallel oriented ellipsoids:

$$f\frac{\hat{\varepsilon}_{\mathrm{me}}(\tilde{\nu}) - \hat{\varepsilon}(\tilde{\nu})}{\hat{\varepsilon}(\tilde{\nu}) + L[\hat{\varepsilon}_{\mathrm{me}}(\tilde{\nu}) - \hat{\varepsilon}(\tilde{\nu})]} = (f - 1)\frac{\hat{\varepsilon}_{\mathrm{po}}(\tilde{\nu}) - \hat{\varepsilon}(\tilde{\nu})}{\hat{\varepsilon}(\tilde{\nu}) + L\left[\hat{\varepsilon}_{\mathrm{po}}(\tilde{\nu}) - \hat{\varepsilon}(\tilde{\nu})\right]}\,.$$

$$(6.25)$$

As for (6.23), in the extension of Bruggeman theory to randomly oriented ellipsoids (6.26) three depolarization factors are introduced.

Bruggeman theory extended to randomly oriented ellipsoids:

$$\frac{\hat{\varepsilon}_{\mathrm{po}}(\tilde{\nu}) - \hat{\varepsilon}(\tilde{\nu})}{\hat{\varepsilon}_{\mathrm{me}}(\tilde{\nu}) - \hat{\varepsilon}_{\mathrm{po}}(\tilde{\nu})} = -\frac{f}{3} \sum_{i=1}^{3} \frac{\hat{\varepsilon}(\tilde{\nu})}{\hat{\varepsilon}(\tilde{\nu}) + L_i\left[\hat{\varepsilon}_{\mathrm{me}}(\tilde{\nu}) - \hat{\varepsilon}(\tilde{\nu})\right]} \, .$$

(6.26)

The effective medium theories discussed up to now result in an exactly determined dielectric function. Another type of model was introduced by Hashin and Shtrikman [561] and further developed by Bergman [562] and Milton [563]. In this approach, bounds were set up for the effective dielectric function. These bounds are valid for extreme nanostructures. The bounds due to Wiener [564] are valid for parallel rods (6.27) perpendicular to the electric field vector and for parallel discs (6.28) parallel to the field vector of the incident light.

Wiener bounds:

$$\hat{\varepsilon}(\tilde{\nu}) = f\hat{\varepsilon}_{\mathrm{me}}(\tilde{\nu}) + (1 - f)\hat{\varepsilon}_{\mathrm{po}}(\tilde{\nu}) \, ,$$

(6.27)

$$\hat{\varepsilon}(\tilde{\nu}) = \frac{\hat{\varepsilon}_{\mathrm{me}}(\tilde{\nu})\hat{\varepsilon}_{\mathrm{po}}(\tilde{\nu})}{\hat{\varepsilon}_{\mathrm{po}}(\tilde{\nu}) + (1 - f)\hat{\varepsilon}_{\mathrm{me}}(\tilde{\nu})} \, .$$

(6.28)

The bounds named after Hashin and Shtrikman [561] are derived for spherical embedded particles. If metal particles are completely embedded in a concentric insulator shell, (6.29) is valid. Equation (6.29) is identical with the Maxwell–Garnett theory (6.21). Otherwise, for the opposite case where insulator particles are embedded in a surrounding metal shell, equation (6.30) is taken. In both cases, the particle can have different sizes, but the ratio between shell thickness and particle radius is constant for all particles.

Hashin–Shtrikman bounds:

$$\frac{\hat{\varepsilon}(\tilde{\nu}) - \hat{\varepsilon}_{\mathrm{po}}(\tilde{\nu})}{\hat{\varepsilon}(\tilde{\nu}) + 2\hat{\varepsilon}_{\mathrm{po}}(\tilde{\nu})} = f\frac{\hat{\varepsilon}_{\mathrm{me}}(\tilde{\nu}) - \hat{\varepsilon}_{\mathrm{po}}(\tilde{\nu})}{\hat{\varepsilon}_{\mathrm{me}}(\tilde{\nu}) + 2\hat{\varepsilon}_{\mathrm{po}}(\tilde{\nu})} \, ,$$

(6.29)

$$\frac{\hat{\varepsilon}(\tilde{\nu}) - \hat{\varepsilon}_{\mathrm{me}}(\tilde{\nu})}{\hat{\varepsilon}(\tilde{\nu}) + 2\hat{\varepsilon}_{\mathrm{me}}(\tilde{\nu})} = (1 - f)\frac{\hat{\varepsilon}_{\mathrm{po}}(\tilde{\nu}) - \hat{\varepsilon}_{\mathrm{me}}(\tilde{\nu})}{\hat{\varepsilon}_{\mathrm{po}}(\tilde{\nu}) + 2\hat{\varepsilon}_{\mathrm{me}}(\tilde{\nu})} \, .$$

(6.30)

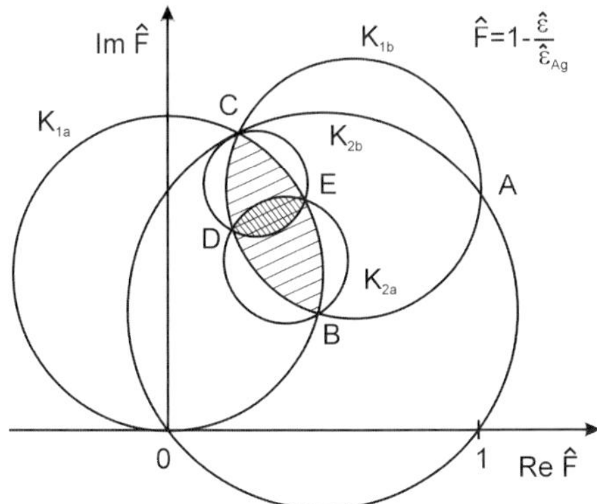

Fig. 6.17. Bounds on the effective dielectric function

As a next step, a function $\hat{F}(\tilde{\nu})$ is defined by

$$\hat{F}(\tilde{\nu}) = 1 - \frac{\hat{\varepsilon}(\tilde{\nu})}{\hat{\varepsilon}_{\mathrm{me}}(\tilde{\nu})} \; .$$

In the complex plane representing its real and imaginary parts, F' and F'', respectively, circles are drawn through the bounds (see Fig. 6.17). The intersection of the circles represents available values for the function $\hat{F}$ at one wave number $\tilde{\nu}$. The intersection of the circle K_{1a}, given by the points calculated from (6.27) and (6.28) together with $\hat{F} = 0$, with the circle K_{1b} calculated from (6.27), (6.28) and $\hat{F}(\tilde{\nu}) = \hat{\varepsilon}_{\mathrm{po}}(\tilde{\nu})$, yields outer bounds on the available values for composite materials in which embedded particles have some tendency towards orientation. All values for $\hat{F}$ calculated with (6.22) for various depolarization factors L are points actually on the circle K_{1b}. The values calculated using (6.25) can be found inside the intersection of the circles K_{1a} and K_{1b}.

Inner bounds arise from the intersection of circles K_{2a} and K_{2b}. These circles were drawn through the values for $\hat{F}$ calculated from (6.27), (6.29) and (6.30), and from (6.28), (6.29) and (6.30), respectively. This intersection represents the available values for embedded particles without any tendency towards orientation. The values found for $\hat{F}$ using (6.23) with different depolarisation factors L_i all lie inside the intersection of circles K_{2a} and K_{2b}.

The Wiener and Hashin–Shtrikman bounds have not been much used for calculating optical properties but the bounds fit very well for approximations to electrical d.c. conductivity [433]. Third-order bounds were introduced in [565] by considering additional parameters that represent the nanostructure of the material.

In the Bergman theory (6.35), the so-called spectral density function $g(x)$ is used to describe the nanostructure of the effective medium. The spectral density function represents the nanostructure of the composite material and includes metal particle size and shape distributions. The definition of the complex $\hat{t}$ function allows the material properties of the metal represented by $\hat{\varepsilon}_{\mathrm{me}}(\tilde{\nu})$ to be discussed independently of the nanostructure. From the $\hat{t}$ function, the spectrally dependent sensitivity $s(\hat{t})$ is given by (6.31). $s(\hat{t})$ describes the sensitivity of the composite system for different nanostructures at various wave numbers:

$$s(\hat{t}) = \begin{cases} -\ln(t'^2 + t''^2) , & t' < 0 , \\ -\ln(t''^2) , & 0 \leq t' \leq 1 , \\ -\ln[(t'-1)^2 + t''^2] , & 1 < t' . \end{cases} \tag{6.31}$$

The spectral density is generally normalized (6.32) and there is an additional condition (6.33) for isotropic effective media [566]:

$$\int_0^1 g(x)\,\mathrm{d}x = 1 , \tag{6.32}$$

$$\int_0^1 xg(x)\,\mathrm{d}x = \frac{1}{3}(1 - f) . \tag{6.33}$$

Bergman theory:

$$\hat{\varepsilon}(\tilde{\nu}) = \hat{\varepsilon}_{\mathrm{po}}(\tilde{\nu}) \left[1 - f \int_0^1 \frac{g(x)}{t(\tilde{\nu}) - x}\,\mathrm{d}x \right] , \tag{6.34}$$

where

$$\hat{t}(\tilde{\nu}) = \frac{\hat{\varepsilon}_{\mathrm{po}}(\tilde{\nu})}{\hat{\varepsilon}_{\mathrm{po}}(\tilde{\nu}) - \hat{\varepsilon}_{\mathrm{me}}(\tilde{\nu})} , \tag{6.35}$$

and $\hat{\varepsilon}_{\mathrm{me}}(\tilde{\nu})$ is the metal dielectric function, $\hat{\varepsilon}_{\mathrm{po}}(\tilde{\nu})$ the insulator dielectric function, $\hat{\varepsilon}(\tilde{\nu})$ the effective dielectric function, f the volume filling factor, and $g(x)$ the spectral density.

If the nanostructure of the composite medium is near the percolation threshold f_{c}, the spectral density function $g(x)$ can be split into a percolating part p and a non-percolating part $g_{\mathrm{r}}(x)$ (6.36):

$$g(x) = p\delta_{\mathrm{r}}(x) + g_{\mathrm{r}}(x) , \quad \text{where} \quad \int_0^n \delta_{\mathrm{r}}(x)\,\mathrm{d}x = 1 \quad \text{for} \quad x > 0 . \tag{6.36}$$

The percolation part p describes the agglomeration of embedded particles as well as the formation of a partly closed (percolated) structure.

The main purpose of the Bergman theory is to take into consideration material and nanostructural properties (particle size and shape distribution). The spectral density function $g(x)$ represents only the nanostructural properties of the composite media. If two composite materials have completely identical nanostructure, the same spectral density function can be used. On the other hand, the main problem with the Bergman theory is the determination of the spectral density function. Substantially different spectral density functions can only be determined for materials with very different nanostructures. As the calculation in Sect. 6.2.4 will show, plasma polymer thin films with embedded silver nanoparticles are just such topologically sensitive systems and these films are well-suited to calculation with the Bergman theory.

There are many other effective medium theories. There are simple formulas which describe the nanostructure without further parameters, due to Looyenga [567], Lichtenecker [568] and Monecke [569]. The effective medium theory by Sheng [570] is valid for particles made from two basic materials. The dipole–dipole interaction between neighboring particles is introduced in the theory due to Persson and Liebsch [571]. Other effective medium theories include multiple scattering [572, 573] and magnetic dipoles [574]. Near-field corrections for ellipsoidal particles have also been included [575]. Effective medium theories have been derived from network calculations [576] or using percolation theories [577]. A generalized effective medium theory was also developed for composite materials with weak non-linearity [578]. All these and other effective medium theories, e.g., [579–581], tend to include additional, sometimes unknown parameters (e.g., for the near-field correction [503]) and require substantial mathematical effort. In addition, some of these models only apply to one specific composite material.

One advance in calculations of the optical properties of composite materials with effective medium theories is their application to a wide range of different materials. Examples are the Maxwell–Garnett theory applied to metal nanoparticles in polymers (Au in polyamide [178]), ceramic–metal composite films (Au–SiO_2 [297], Pt–Al_2O_3 [222], Co–Al_2O_3 [582]), Ag, Au, or Al nanoparticles in nanoporous alumina [583, 584], and dyes on silver surfaces [585]. The Bruggeman theory has been applied to ceramic–metal composite films such as Pt–Al_2O_3 [222]), multilayer structures made from In-GaP/InGaAs [586], partly amorphous GaAs [587], and copper oxide growth [588]. Both theories have been used in combination to describe the optical properties of Au/Ag nanoalloys in nanoporous alumina [589]. The limiting bounds were used to describe Co–Al_2O_3 composite films, e.g., [221]. All the results of these calculations concerning optical properties have given reasonable agreement with experimental data. It is clear that the Maxwell–Garnett and Bruggeman theories, and particularly their extensions to ellipsoidal particles, are universally applicable approximations which can be used for a large number of composite materials. An exact general description of optical properties as a function of nanostructure is only possible for some special composite materials. In the following, it will be demonstrated that calculations with effective medium theories can be successfully applied to plasma polymer films with embedded silver nanoparticles.

6.2.4 Calculational Results

Assessment. Since transmission spectra are the simplest way to describe optical properties, the comparison here between experimental and modelled optical properties is made using experimental and modelled transmission spectra. To begin with, criteria for successful modelling have to be formulated. The following criteria were used to assess models for plasma polymer films with embedded metal nanoparticles:

- correct effective dielectric functions ($\varepsilon''(\tilde{\nu}) \geq 0$),
- spectral position and shape of transmission minimum,
- spectral shift of transmission minimum to higher wave numbers with decreasing filling factor,
- spectral shift of transmission minimum to higher wave numbers with decreasing filling factor as a result of thermal treatment,
- quantitative spectral behavior.

Concerning further criteria, when calculating transmission spectra for multilayer systems with embedded metal particles, the total thickness d and thicknesses of each layer d_{p1}, d_{c} and d_{p2} must be considered in the right way. Otherwise the concurrence between the actual nanostructure and the particle geometry used in the effective medium theory has to be demonstrated. This means that knowledge of the nanostructure obtained from TEM micrographs has be taken into account in calculations. In the following, results from particle size and shape analysis were transferred to the parameters used in effective medium theories. These were the volume filling factor f, the depolarization factors L or L_i, and the composite film thickness d_{c}.

From image analysis of lateral TEM micrographs, the area filling factor F_{A} and mean values $\bar{D}$ and $\bar{S}$ for the particle size and shape are given. The maximum mean particle diameter $\bar{D}_{\mathrm{max}}$ is defined by $\bar{D}_{\mathrm{max}} = \bar{D} + \sigma_D$. The area filling factor F_{A} only agrees with the volume filling factor f if the particles have prism shape with ellipsoidal base area. In this case $f = F_{\mathrm{A}}$ and the height of all prisms is equal to the composite film thickness d_{c} (see Fig. 6.18).

Cross-sectional TEM investigations (see Sect. 3.2.2) showed that vertical projections of the particles are also ellipsoids, with diameters D_A, D_B, D_C. It is useful to assume that the two lowest diameters are identical, so that $D_B = D_C$ and there is rotational symmetry. The values of $\bar{D}_A$ and $\bar{D}_B$ can be calculated from $\bar{D}$ and $\bar{S}$ using $\bar{D}_A = \bar{D}(1 + \sqrt{1 - \bar{S}})$ and $\bar{D}_B = \bar{D}(1 - \sqrt{1 - \bar{S}})$.

By comparing the ratio of the prism volume to the ellipsoid volume, a maximal volume filling factor f_{max} follows from the area filling factor using $f_{\mathrm{max}} = 2F_{\mathrm{A}}/3$. This upper bound on the volume filling factor is valid if all particles have an identical size $\bar{D}_{\mathrm{max}}$. In this case, the minimum composite film thickness corresponds to the mean vertical diameter of the particles $d_{\mathrm{c}}^{\mathrm{min}} = \bar{D}_B$ (provided that all particles lie exactly in one plane). Because there is a statistical particle size distribution, the composite film thickness must be defined from the maximal vertical diameter of the largest particle $d_\mathrm{c}^\mathrm{max} = D_B^\mathrm{max}$. The value of D_B^max can be obtained from image analysis data. It is better to calculate D_B^max using $D_B^\mathrm{max} = D_\mathrm{max}(1 - \sqrt{1 - S_\mathrm{max}})$. The filling factor f is further decreased by the ratio between the mean particle diameter $\bar{D}_B$ and the maximum vertical particle diamet er D_B^max so that

$$ f_\mathrm{min} = \frac{\bar{D}_B}{D_B^\mathrm{max}} \frac{2}{3} F_\mathrm{A} \; . $$

On the basis of these geometrical assumptions, the lower and upper bounds $f_\mathrm{min} \le f \le f_\mathrm{max}$ act as further criteria for the success of calculations with effective medium theories.

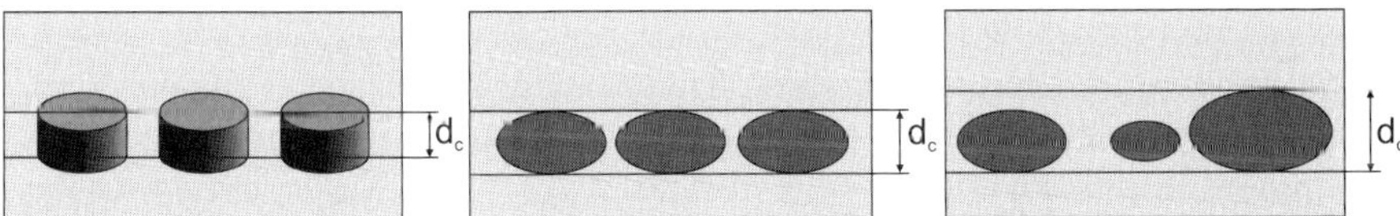

Fig. 6.18. Estimating the volume filling factor f for embedded nanoparticles. *Left:* $f = F_\mathrm{A}$. *Centre:* $f_\mathrm{max} = 2F_\mathrm{A}/3$. *Right:* $f_\mathrm{min} = \dfrac{\bar{D}_B}{D_B^\mathrm{max}} \dfrac{2}{3} F_\mathrm{A}$

The depolarization factor L in (6.22) and (6.25) and the depolarization factors L_i in (6.23) and (6.26) can be calculated from the mean aspect ratio $\bar{Q}$ and the mean shape factor $\bar{S}$ of the particles. For ellipsoids with diameters $\bar{D}_A$, $\bar{D}_B$, $\bar{D}_C$, the factors L_i are given by [556]

$$ L_i = \frac{\bar{D}_A \bar{D}_B \bar{D}_C}{16} \int \frac{\mathrm{d}u}{(\bar{D}_i + 4u)\sqrt{(\bar{D}_A^2 + 4u)(\bar{D}_B^2 + 4u)(\bar{D}_C^2 + 4u)}} \, . \tag{6.37} $$

Using (6.37) and $\sum L_i = 1$, the depolarization factors L and L_i can be calculated. In the following, only one depolarization factor L is used. In the left-hand part of Fig. 6.19, L is given as a function of the shape factor S. The lower curve is valid for rods ($L < 1/3$) and the upper curve for discs ($L > 1/3$). The depolarization factors L_i for rods ($L_1 > L_2 = L_3$) and for plates ($L_1 < L_2 = L_3$) are also given in Fig. 6.19. For rods, the depolarization factor L_1 is given from the upper curve, whilst $L_2 = L_3$ are given from the lower curve in the left-hand part of Fig. 6.19. The right-hand part of Fig. 6.19 is valid for discs. L_1 is given from the lower curve and $L_2 = L_3$ from the upper curve.

Maxwell–Garnett Theory and Bruggeman Theory. We begin by calculating[1] the optical properties of plasma polymer films with embedded silver

[1] Fsos software package 1986,1992,1996, TU Chemnitz/Fraunhofer Institute for Material and Beam Technology.

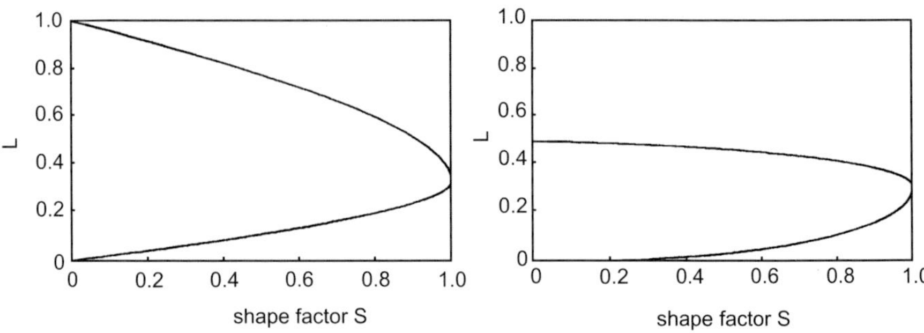

Fig. 6.19. Depolarisation factors L and L_i as function of the shape factor S for various particle geometries

nanoparticles using the Maxwell–Garnett and Bruggeman theories without an experimentally determined film thickness and without considering any limitations for the filling factor value. Calculations are then given for an experimental multilayer system including a plasma polymer layer with embedded nanoparticles.

Figure 6.20 shows the real and imaginary parts of the effective dielectric function as well as transmission spectra resulting from calculations with the Maxwell–Garnett theory (left) and the Bruggeman theory (right). Maxwell–Garnett calculations for effective dielectric functions $\hat{\varepsilon}(\tilde{\nu})$ are done with the dielectric functions of silver $\hat{\varepsilon}_{\mathrm{Ag}}(\tilde{\nu})$ and the plasma polymer $\hat{\varepsilon}_{\mathrm{pp}}(\tilde{\nu})$, both given in Fig. 6.13, with filling factors (a) $f = 0.1$, (b) $f = 0.2$, (c) $f = 0.3$ and (d) $f = 0.4$. The resonance shifts to lower wave numbers with increasing filling factor while the intensity and the half-width increase. From these effective dielectric functions, transmission spectra were calculated for a multilayer system plasma polymer/plasma polymer with embedded silver nanoparticles/plasma polymer with thickness $d_{\mathrm{p1}} = 40$ nm, $d_{\mathrm{c}} = 20$ nm and $d_{\mathrm{p2}} = 40$ nm. There is a shift to lower wave numbers with incre asing filling factor while the intensity of the plasma resonance absorption increases. The constant transmission point found in experiments at $\tilde{\nu} = 26\,000$ cm^{-1} (see Figs. 6.1 and 6.5) is also visible in the calculated spectra.

The conditions for the occurrence of this constant transmission point and its spectral position will be discussed in Sect. 6.2.4. In conclusion, calculations with the Maxwell–Garnett theory give the following positive results:

- observed spectral shift of the plasma resonance absorption to lower wavelengths with increasing filling factor or particle size,
- observed increase in intensity of the plasma resonance absorption with increasing particle size,
- good correspondence between the shape of the plasma resonance absorption (Lorentz peak) and experimental data,
- occurrence of a constant transmission point.

There are two points where they are less successful:

- intensities of the plasma resonance absorption are too high and the half-widths are too low,
- the calculated spectral position of the plasma resonance absorption only corresponds to the experimental position if further limiting assumptions are made.

The usual result of calculations with the Maxwell–Garnett theory is a good correspondence with spectral position but over-high intensities or excessively low half-widths, e.g., for Ag particles in gelatin [573], Ag–SiO$_2$ composite films [197,482,590], Ag–MgF$_2$ composite films [591,592], ion implanted Ag nanoparticles in glass [593], plasma polymer–gold composite films (Au–C$_3$F$_8$ [256,258], Au–C$_2$F$_3$Cl [276]) and thin metal layers (Ag [494], Au [347]). One way of reducing the intensity of the plasma resonance absorption is to introduce additional damping factors in (6.21) [494] or to use a dielectric function calculated from the Drude theory using a damping factor. The Maxwell–Garnett theory has also been used to extract nanostructural data from optical constants determined using variable angle spectroscopic ellipsometry (Au in plasma polymer made from C$_4$F$_8$) [268].

The calculation with the Maxwell–Garnett theory for parallel oriented ellipsoids (6.22) takes particle shape into account and results in a spectral shift of the plasma resonance absorption which depends on the depolarization factor. The results will be given later.

The calculation with the Maxwell–Garnett theory for random oriented ellipsoids (6.23) results in a splitting of the plasma resonance absorption into two resonances with the same intensity [433]. This does not agree with experimental data for particle assemblies.

Calculations with the Bruggeman theory, shown on the right in Fig. 6.20, lead to completely different results. The effective dielectric functions $\hat{\varepsilon}(\tilde{\nu})$ were calculated with the dielectric functions of silver $\hat{\varepsilon}_{Ag}(\tilde{\nu})$ and the plasma polymer $\hat{\varepsilon}_{po}(\tilde{\nu})$ from Fig. 6.13 using (6.24) and filling factors (a) $f = 0.05$, (b) $f = 0.1$, (c) $f = 0.15$ and (d) $f = 0.2$. The very broad resonance shifts only very slightly to lower wave numbers with increasing filling factors. For higher filling factors, the resonances are also broadened to lower wave numbers and they are no longer found in the spectral region shown in Fig. 6.20. While the intensity of the plasma resonance absorption is much lower than that calculated with the Maxwell–Garnett theory, calculation of the transmission spectra for a plasma polymer film with embedded silver nanoparticles was done using a film thickness of $d = 100$ nm. The results of calculations give plasma resonance absorptions which are very broad and shift only slightly with increasing filling factor. In conclusion, the Bruggeman theory has only one successful result:

- increase in intensity of the plasma resonance absorption with increasing filling factor or particle size.

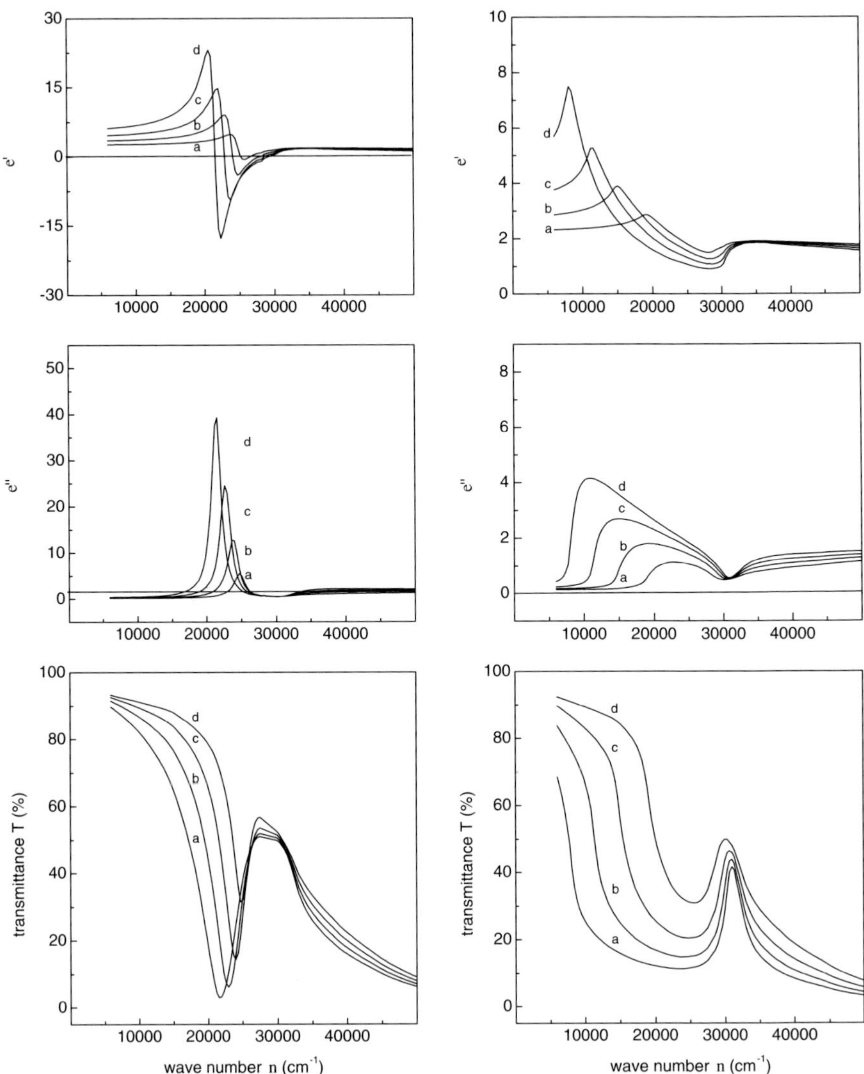

Fig. 6.20. Effective dielectric functions and transmission spectra calculated with the Maxwell–Garnett theory (*left*) and the Bruggeman theory (*right*). Maxwell–Garnett theory: (a) $f = 0.1$, (b) $f = 0.2$, (c) $f = 0.3$ and (d) $f = 0.4$, $d_{p1} = 40$ nm, $d_c = 20$ nm, $d_{p2} = 40$ nm. Bruggeman theory: (a) $f = 0.05$, (b) $f = 0.1$, (c) $f = 0.15$ and (d) $f = 0.2$, $d = 100$ nm

Negative points are:

- no spectral shift in the plasma resonance absorption to lower wave numbers with increasing particle size,
- the shape of the plasma resonance absorption does not compare well with experimental results,
- no constant transmission point,
- half-widths of the plasma resonance absorption are too large,
- the calculated experimental position of the plasma resonance absorption only corresponds to experimental data in a few cases.

It becomes obvious that calculations with the Bruggeman theory are not very successful. Using the Bruggeman theory for parallel oriented ellipsoids (6.25) and various depolarization factors, a spectral shift of the plasma resonance absorption can be calculated but all the other imperfections of the calculation are still present. Consequently, no further calculations with Bruggeman theory are presented.

The calculation with the Maxwell Garnett theory for parallel oriented ellipsoids (6.22) was fitted to experimental transmission spectra for a multilayer system with embedded silver nanoparticles before and after thermal treatment (Fig. 6.5). Furthermore, the particle size and shape distributions determined from TEM micrographs (Figs. 4.5c–h and 4.6c–h) for a comparable multilayer system were included in the calculations.

Taking into account the considerations about the volume filling factors f given at the beginning of Sect. 6.2.4, the thickness d_c of the particle-containing layer of the multilayer system and the depolarization factors L given from Fig. 6.19, the nanostructural values for the following calculations are as follows.

Before thermal treatment:

(c)	$0.32 \leq f \leq 0.60$	$0.10 \leq L \leq 0.23$	$11 \leq d_c \leq 21$ nm
(e)	$0.31 \leq f \leq 0.51$	$0.14 \leq L \leq 0.24$	$10 \leq d_c \leq 17$ nm
(g)	$0.26 \leq f \leq 0.49$	$0.11 \leq L \leq 0.25$	$7 \leq d_c \leq 17$ nm

After thermal treatment:

(d)	$0.25 \leq f \leq 0.49$	$0.13 \leq L \leq 0.23$	$14 \leq d_c \leq 29$ nm
(f)	$0.20 \leq f \leq 0.44$	$0.14 \leq L \leq 0.24$	$12 \leq d_c \leq 22$ nm
(h)	$0.24 \leq f \leq 0.40$	$0.16 \leq L \leq 0.25$	$10 \leq d_c \leq 18$ nm

The thicknesses d_{p1} and d_{p2} are determined in such a way as to yield a constant total thickness $d = d_{p1} + d_c + d_{p2} = 100$ nm.

Because the intensity of the transmission minima is too high and the half-width too low, the last criterion pronounced in Sect. 6.2.4, which refers to quantitative spectral behavior, is not satisfied. On the other hand, the existence of a constant transmission point before and after thermal treatment can be added to the criteria. In earlier calculations with (6.22), it was found that transmission spectra calculated with various filling factors only show a constant transmission if the depolarization factor L and the thickness of the composite film d_c remain constant.

The calculations were done in such a way that the experimental position of the constant transmission point fitted the spectra given in Fig. 6.5. In Sect. 6.5, the constant transmission point was determined at $\tilde{\nu} = 22\,000$ cm^{-1} before thermal treatment and $\tilde{\nu} = 23\,000$ cm^{-1} after thermal treatment. The constant transmission point shifts to higher wave numbers with increasing depolarization factor L [433]. For this reason the depolarization factors $L = 0.2$ and $L = 0.25$ were used for calculations before and after thermal treatment. This represents experimental data for the particle size and shape distribution in such a way that the mean shape factor of the particles, which is coupled to the depolarization factor, increases with thermal treatment.

To save the constant transmission point during the calculations, the thickness of the composite film d_c also has to be constant for the spectra before and after thermal treatment, respectively. Following the results for the particle size and shape distributions, thicknesses for the composite films are deduced with $d_c = 25$ nm before thermal treatment and $d_c^{\text{th}} = 30$ nm after thermal treatment. To preserve a constant total film thickness $d = 100$ nm, the increase in the composite film thickness requires a reduction in the thicknesses d_{p1} and d_{p2} of the two plasma polymer films on either side.

As thermal treatment changes the particle size and shape distribution without mass loss, for spectral calculations after thermal treatment the filling factor has to recalculated from the composite film thickness using $f^{\text{th}} = d_c/d_c^{\text{th}}$. The filling factor f and depolarization factor L used for the calculation in Fig. 6.21 lie inside the range determined at the beginning of this chapter. The composite film thickness d_c is situated in the upper region of the given range.

Figure 6.21 demonstrates that, using the described assumptions, it is possible to reproduce the experimentally observed shift in the transmission minima to higher wave number during thermal treatment. Moreover, the increase in transmission values for the constant transmission point is also calculated. It can be observed that changes in optical transmission during thermal treatment are reproduced qualitatively.

It should be noted that the extensive information about particle size and shape distributions from the optical image analysis of TEM micrographs can be used only partially in calculations with the Maxwell–Garnett theory and its extensions. A sufficient calculation requires assumptions which bypass or neglect nanostructural information. Otherwise the simple Maxwell–Garnett theory for parallel oriented ellipsoids gives a good qualitative description of transmission spectra and their changes after thermal treatment.

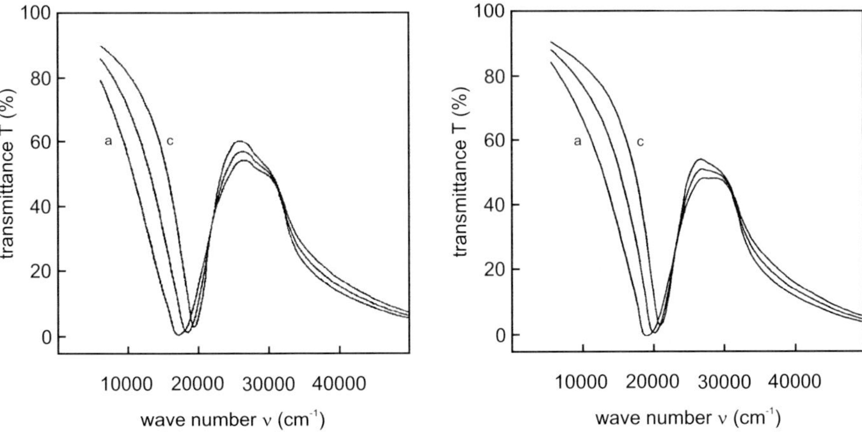

Fig. 6.21. Calculated transmission spectra (Maxwell–Garnett theory for parallel oriented ellipsoids) for a multilayer system with embedded silver nanoparticles (*left*) before and (*right*) after thermal treatment (Fig. 6.5). *Left*: (a) $f = 0.45$, (b) $f = 0.35$, (c) $f = 0.25$; $L = 0.20$, $d_{p1} = 40$ nm, $d_c = 25$ nm, $d_{p2} = 35$ nm. *Right*: (a) $f = 0.38$, (b) $f = 0.29$, (c) $f = 0.21$; $L = 0.25$, $d_{p1} = 35$ nm, $d_c = 30$ nm, $d_{p2} = 35$ nm

Bergman Theory. For calculations with the Bergman theory, it is necessary to model the main dependence of optical properties on the nanostructure of the material. For this purpose, the $\hat{t}$ function defined in (6.35) and the spectral sensitivity $s(t)$ from (6.31) were used. Figure 6.22 shows the $\hat{t}$ function and spectral sensitivity $s(t)$ for plasma polymer thin films with embedded silver particles. Once again, the dielectric function used was $\hat{\varepsilon}_{Ag}(\tilde{\nu})$ as determined experimentally by Johnson and Christy. Experimental values from Fig. 6.13 were taken for the plasma polymer dielectric function $\hat{\varepsilon}_{po}(\tilde{\nu})$. In the plane spanned by t' and t'', the values for $\hat{t}$ cover a wide range. The scope for using different spectral densities $g(x)$ for various nanostructures is great

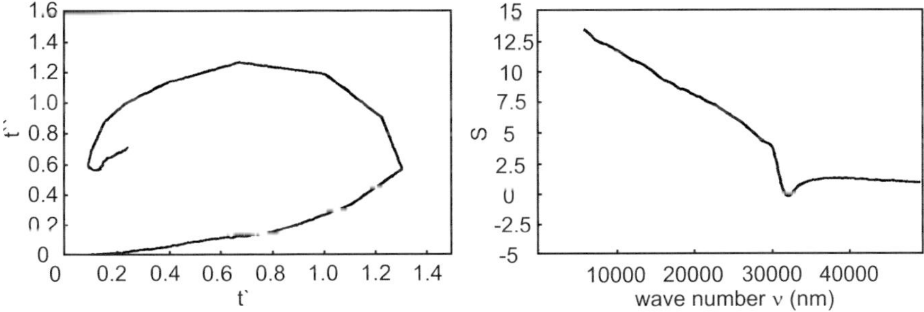

Fig. 6.22. *Left*: complex t function (t'' vs. t') and *right*: spectral sensitivity $s(t)$ for plasma polymer thin films with embedded silver (*dashed line*) and gold (*dotted line*) nanoparticles

for plasma polymer thin films with embedded silver particles. The spectral sensitivity of silver already reaches values $s(\hat{t}) > 1$ at wave numbers below $\tilde{\nu} = 30\,000$ cm^{-1}. This demonstrates that in the region of plasma resonance absorption, between $10\,000 \leq \tilde{\nu} \leq 25\,000$ cm^{-1}, the spectral density has a substantial influence on the calculated spectra.

Following these introductory remarks concerning the $\hat{t}$ function and spectral sensitivity $s(\hat{t})$, the results of Bergman calculations are now given for plasma polymer films with embedded silver nanoparticles. In calculations with (6.35), the spectral density $g(x)$ acts as a fit function and experimental and calculated spectra agree well.

The left-hand part of Fig. 6.23 shows the experimental and calculated transmission spectra for a multilayer system with embedded silver particles before and after thermal treatment (film II/3). Once again, the dielectric functions from Fig. 6.13 were used. The right-hand part of Fig. 6.23 gives the spectral densities $g(x)$ used for the calculations. The best fit for the calculated spectra before thermal treatment was obtained using a spectral density $g(x)$ which includes a so-called percolation part $(g(0) > 0)$. This implies that nanoparticles are not well-separated. After thermal treatment, the spectral density $g(x)$ does not have a percolating part for the best fit. This agrees with the observed nanostructure of the particles before and after thermal treatment, e.g., in Fig. 4.5.

A detailed description of calculations with the Bergman theory is given in [594, 595]. Calculations have been presented for silver particles [596], bipolyethylene composites [597] and Pt–Al$_2$O$_3$ composite materials [598]. The spectral densities arising in these calculations cannot be used in calculations for plasma polymer–silver composite films. If the same spectral densities were used, both materials would have to have the same topological structure or,

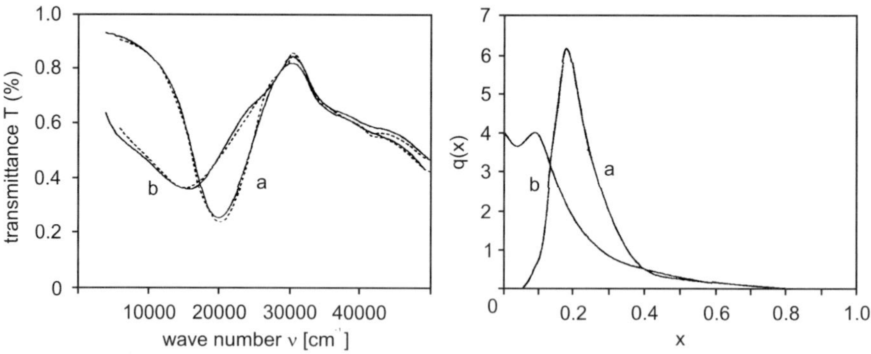

Fig. 6.23. Transmission spectra calculated using the Bergman theory and spectral densities for a plasma polymer film with embedded silver nanoparticles. *Left*: experimental (*continuous curve*) and calculated (*dotted curve*) transmission spectra before and after thermal treatment. *Right*: spectral densities $g(x)$ for calculated transmission spectra [594]

in other words, the same particle size and shape distributions. It is proposed that the spectral density found for plasma polymer films with embedded silver nanoparticles can also be used for films with embedded gold or copper nanoparticles if these films have exactly the same particle size and shape distributions. The main disadvantage with the Bergman theory is that the spectral density is not coupled to a nanostructure that is valid for different nanostructured materials.

Constant Transmission Point. The experimentally determined constant transmission point (Figs. 6.1, 6.5 and 6.9) was also found when calculating optical spectra with the Maxwell–Garnett theory (Fig. 6.20). In the following, the mathematical genesis of this constant transmission point will be described.

Equation (6.21) expressing the Maxwell–Garnett theory can be rewritten in the form

$$\frac{\hat{\varepsilon}(\tilde{\nu}) - \hat{\varepsilon}_{\mathrm{po}}(\tilde{\nu})}{f[\hat{\varepsilon}(\tilde{\nu}) + 2\hat{\varepsilon}_{\mathrm{po}}(\tilde{\nu})]} = \frac{\hat{\varepsilon}_{\mathrm{me}}(\tilde{\nu}) - \hat{\varepsilon}_{\mathrm{po}}(\tilde{\nu})}{\hat{\varepsilon}_{\mathrm{me}}(\tilde{\nu}) + 2\hat{\varepsilon}_{\mathrm{po}}(\tilde{\nu})} \ . \tag{6.38}$$

The right-hand side of (6.38) does not depend on the effective dielectric constant and remains constant for a given wave number, so that

$$\frac{\hat{\varepsilon}(\tilde{\nu}) - \hat{\varepsilon}_{\mathrm{po}}(\tilde{\nu})}{f[\hat{\varepsilon}(\tilde{\nu}) + 2\hat{\varepsilon}_{\mathrm{po}}(\tilde{\nu})]} = \mathrm{Const.}\,(\tilde{\nu}) \ . \tag{6.39}$$

If the denominator in (6.39) becomes very small,

$$\hat{\varepsilon}(\tilde{\nu}) + 2\hat{\varepsilon}_{\mathrm{po}}(\tilde{\nu}) \longrightarrow 0 \ , \tag{6.40}$$

the multiplication by the filling factor can be neglected and

$$\hat{\varepsilon}(\tilde{\nu}) \approx -2\hat{\varepsilon}_{\mathrm{po}}(\tilde{\nu}) \ . \tag{6.41}$$

This implies that the effective dielectric function $\hat{\varepsilon}(\tilde{\nu})$ does not depend on the filling factor. If (6.38) is independent of the filling factor for $\hat{\varepsilon}(\tilde{\nu}) \approx -2\hat{\varepsilon}_{\mathrm{po}}(\tilde{\nu})$, the real part of the effective dielectric function is constant at $\varepsilon'(\tilde{\nu}) = \varepsilon'_{\mathrm{po}}(\tilde{\nu})$ provided that the imaginary part is very small $\varepsilon''(\tilde{\nu}) \approx 0$.

This is illustrated in Fig. 6.24 on the basis of calculations using the Maxwell–Garnett theory for various filling factors. To avoid the influence of optical absorption by the plasma polymer, a constant dielectric function was used, viz., $\varepsilon'_{\mathrm{po}}(\tilde{\nu}) = 2.8$ and $\varepsilon''_{\mathrm{po}}(\tilde{\nu}) = 0$. Once again, the values for the dielectric function were taken from Fig. 6.13 after Johnson and Christy. A constant value for the real part of the effective dielectric function was calculated at $\varepsilon'(\tilde{\nu}) = -5.6$, exactly twice the real part of the dielectric function of the plasma polymer. At wave number $\nu = 23\,500$ cm^{-1}, the imaginary part of the dielectric function is still $\varepsilon''(\tilde{\nu}) \approx 0$.

Figure 6.25 gives transmission spectra calculated using the Maxwell–Garnett theory for filling factors $f = 0.05, f = 0.1, f = 0.15, \ldots, f = 1.0$ and two different film thickness. All transmission spectra intersect in one point at

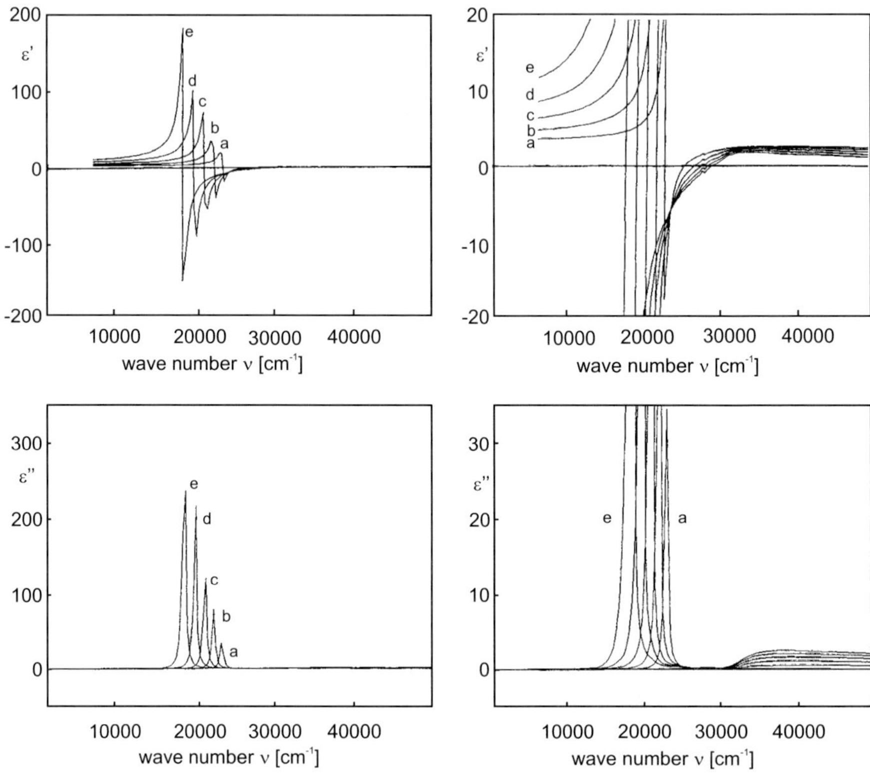

Fig. 6.24. *Left*: effective dielectric functions (Maxwell–Garnett theory). *Right*: section enlargement (a) $f = 0.1$, (b) $f = 0.2$, (c) $f = 0.3$, (d) $f = 0.4$, and (e) $f = 0.5$

$\tilde{\nu} = 23\,500$ cm^{-1}. It is obvious that the occurrence of constant transmission is a direct consequence of the relation $\hat{\varepsilon}(\tilde{\nu}) = -2\hat{\varepsilon}_{\mathrm{po}}(\tilde{\nu})$. The transmission value for the constant transmission point decreases with increasing thickness.

The fact that constant transmission point results from the relation $\varepsilon'(\tilde{\nu}) = -2\varepsilon'_{\mathrm{po}}(\tilde{\nu})$ in the Maxwell–Garnett theory should help us to understand the origin of (6.21). Equation (6.21) follows simply from the Clausius–Mossotti equation

$$\frac{\hat{\varepsilon}(\tilde{\nu}) - \hat{\varepsilon}_{\mathrm{po}}(\tilde{\nu})}{\hat{\varepsilon}(\tilde{\nu}) + 2\hat{\varepsilon}_{\mathrm{po}}(\tilde{\nu})} = \frac{1}{3\varepsilon_0} \sum n_i \alpha_i \, , \tag{6.42}$$

by inserting the polarizability α of (6.4). This is possible under the assumption that the polarizability α is the same for all particles. The volume filling factor is given by [599]

$$f = \rho_i \frac{4}{3} \pi R^3 \, ,$$

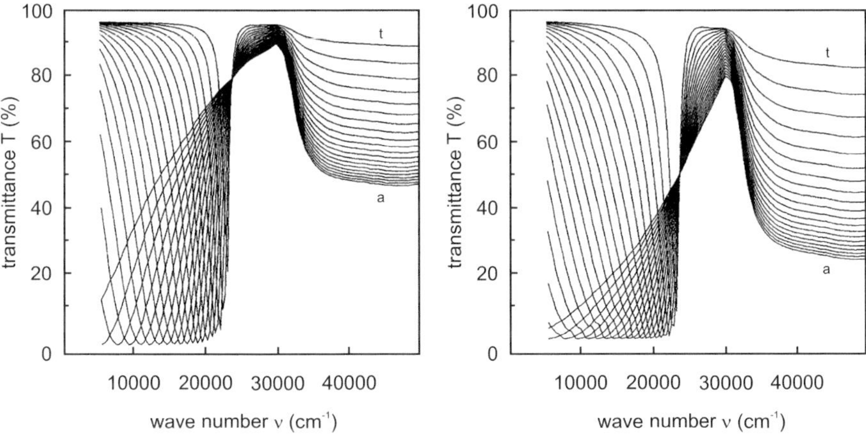

Fig. 6.25. Calculated transmission spectra (Maxwell–Garnett theory) for two plasma polymer films with embedded silver nanoparticles for filling factors $f = 0.05, 0.1, 0.15, \ldots, 1$, film thickness $d_c = 10$ nm (*left*) and $d_c = 20$ nm (*right*)

where ρ_i is the mass density. The left-hand sides of (6.21) and (6.42) are identical. It follows that, at the constant transmission point, the effective dielectric function $\hat{\varepsilon}(\tilde{\nu})$ does not depend on particle size.

The Clausius–Mossotti equation derives directly from the Lorentz field assumption for cubic crystals. The model describes the local electric field $\boldsymbol{E}_{\mathrm{loc}}$ in the neighbourhood of a sphere cut from the dielectric medium, viz.,

$$\boldsymbol{E}_{\mathrm{loc}} = \boldsymbol{E} + \boldsymbol{E}_2 , \quad \boldsymbol{E}_2 = \frac{1}{3\varepsilon_0}\boldsymbol{P} . \tag{6.43}$$

In (6.43), $\boldsymbol{E}$ is the macroscopic field outside the sphere, $\boldsymbol{E}_2$ the Lorentz field resulting from the polarization on the sphere surface and $\boldsymbol{P}$ the polarization. The Lorentz field $\boldsymbol{E}_2$ is calculated from the integral of the surface charge density

$$\boldsymbol{E}_2 = -\frac{1}{4\pi\varepsilon_0 r^2} \int_S \boldsymbol{P} \cdot \frac{\boldsymbol{r}}{r} \, \mathrm{d}^2 r \tag{6.44}$$

over the sphere surface S. It can be stated that the constant transmission point results directly from this Lorentz field assumption, valid for particles larger than about 1 000 atoms.

From these considerations the constant transmission point should be found in all polymer films with embedded silver particles and also in other embedding media if the dielectric function $\varepsilon_{\mathrm{ins}}(\tilde{\nu})$ of the host medium fulfills the condition $\varepsilon(\tilde{\nu}) = -2\varepsilon_{\mathrm{ins}}(\tilde{\nu})$. First of all, the constant transmission point was found in plasma polymer multilayers with embedded silver particles that were mainly deposited by alternating plasma polymerization and metal evaporation (Fig. 6.1).

The occurrence of the constant transmission point for various deposition properties of the plasma polymer film was proved by a large number of optical measurements. The thickness d of the multilayer system can be much higher $d > 100$ nm. In addition, the thicknesses $d_{p1} \leq d_{p2}$, $d_{p1} = d_{p2}$ or $d_{p1} \geq d_{p2}$ of the two embedding polymer films does hinder the occurrence of a constant transmission point if the thickness is constant for various metal contents. Furthermore, plasma polymer thin films have be to deposited with low absorption in the spectral region $20\,000 \leq \tilde{\nu} \leq 25\,000$ cm^{-1} (Fig. 6.1) and the first plasma polymer film, which is the substrate for the evaporated particles, must have low surface roughness. The constant transmission point can also be found after thermal treatment (Fig. 6.5), even if the multilayer system is destroyed (Fig. 6.9).

The investigation of nanostructure given in Chap. 3 does not give a criterion to connect the occurrence of a constant transmission point with a specific silver particle size and shape distribution. There is also no connection with the cauliflower morphology of the plasma polymer.

On the basis of experimental observations, it can be stated that the following conditions have to be realized to observe the constant transmission point. Firstly, there must be a homogeneous matrix around the silver particles so that the integral of the surface charge can be applied. Secondly, the matrix around the silver particles has to have a unique thickness.

Up to now, the constant transmission point has only been found in plasma polymer films with embedded silver particles. One reason could be that the measurement must be made on films with continuously varying filling factor. It is then possible to measure a large number of spectra with various metal particle size and shape distributions but with constant polymer properties in just one film. Optical measurements of silver particles embedded in various embedding media have often been made on various films deposited separately. However, in the particular case of highly absorbent media, tiny changes in the film thickness can change the transmission value and a constant transmission point does not occur. Moreover, for evaporated discontinuous silver films, the constant transmission point is not observed because there are two different media surrounding the nanoparticles (vacuum or air and a substrate material such as quartz).

Considering the experimental data and the given explanation, the constant transmission point can only be observed for silver nanoparticles with various particle size and shape distributions if they are embedded in homogeneous low-absorbing dielectric media with constant film thickness. The occurrence of a constant transmission point for thin insulating films with embedded silver nanoparticles can be understood by a Lorentz field description within classical electromagnetic theory. The assumptions made for spherical particles with the same polarizability can also be used for non-spherical embedded particles with statistical size and shape distributions. Nanostructural information is difficult to treat in calculations with the electromagnetic theory. However, it has been demonstrated that assumptions made in the context of the electromagnetic theory are valid for particles with a wide range of nanostructures.

6.3 Correlation of Nanostructure with Optical Properties

In the following, nanostructure and optical properties will be correlated on the basis of experimental results and calculations.

Figures 6.26 and 6.27 show TEM micrographs and the three-dimensional particle size and shape histograms of two silver nanoparticle assemblies embedded in the same plasma polymer thin film (film I/5). The figure also gives the measured and calculated transmission spectra (Bergman theory), and the spectral densities. The TEM micrographs and histograms correspond to Figs. 3.8 and 3.9. For the samples from Figs. 6.26 and 6.27, the mean particle

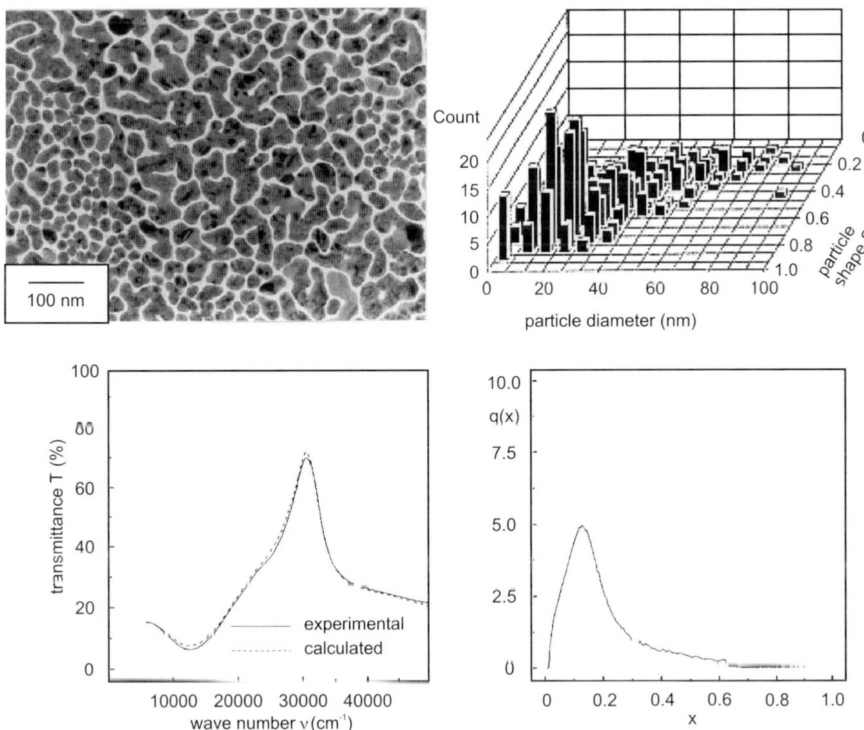

Fig. 6.26. Nanostructure and optical properties of plasma polymer films with embedded silver nanoparticles. *Top left*: TEM micrograph. *Top right*: three-dimensional particle size and shape histogram. *Bottom left*: experimental and calculated transmission spectra. *Bottom right*: spectral density (Bergman theory)

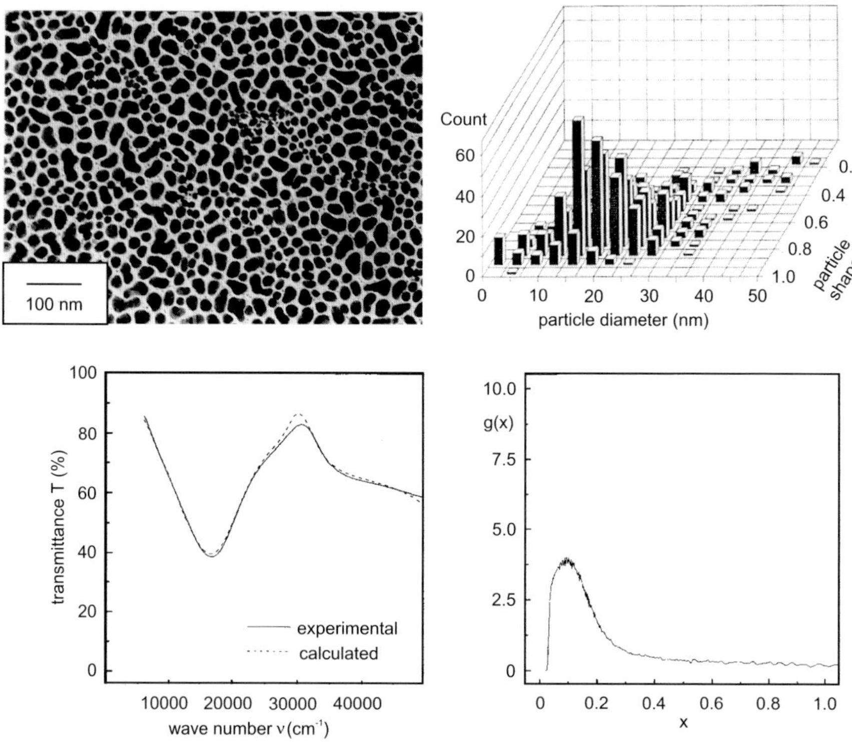

Fig. 6.27. Nanostructure and optical properties of plasma polymer films with embedded silver nanoparticles. *Top left*: TEM micrograph. *Top right*: three-dimensional particle size and shape histogram. *Bottom left*: experimental and calculated transmission spectra. *Bottom right*: spectral density (Bergman theory)

size and mean shape factors have the following values: $\bar{D} = 30.0 \pm 19.5$ nm, $\bar{S} = 0.56 \pm 0.20$ and $\bar{D} = 20.8 \pm 8.0$ nm, $\bar{S} = 0.70 \pm 0.16$, respectively. The transmission minima (extinction maxima) are found at $\tilde{\nu}_r = 13\,000$ cm^{-1} and $\tilde{\nu}_r = 17\,000$ cm^{-1}, respectively. The half-width of the transmission minimum is $\Gamma = 13\,500$ cm^{-1} and the relative intensity is $I_m = 1.32$ (Fig. 6.26).

Figure 6.28 shows the correlation between the optical plasma resonance absorption and the mean particle size and shape for a plasma polymer multi-layer with embedded silver nanoparticles (film I/20). With increasing particle size, there is a red shift of the plasma resonance absorption and an increase in its mean half-width. With increasing shape factor, which means that the particles become more spherical, the plasma resonance absorption shifts to higher wave numbers. The half-width Γ also varies with the mean particle diameter and mean shape factor. It decreases with decreasing mean particle size and with decreasing deviations of the particles from the spherical shape. Figure 6.28 shows very clearly that particle size and shape must be considered when describing optical properties of non-spherical particles.

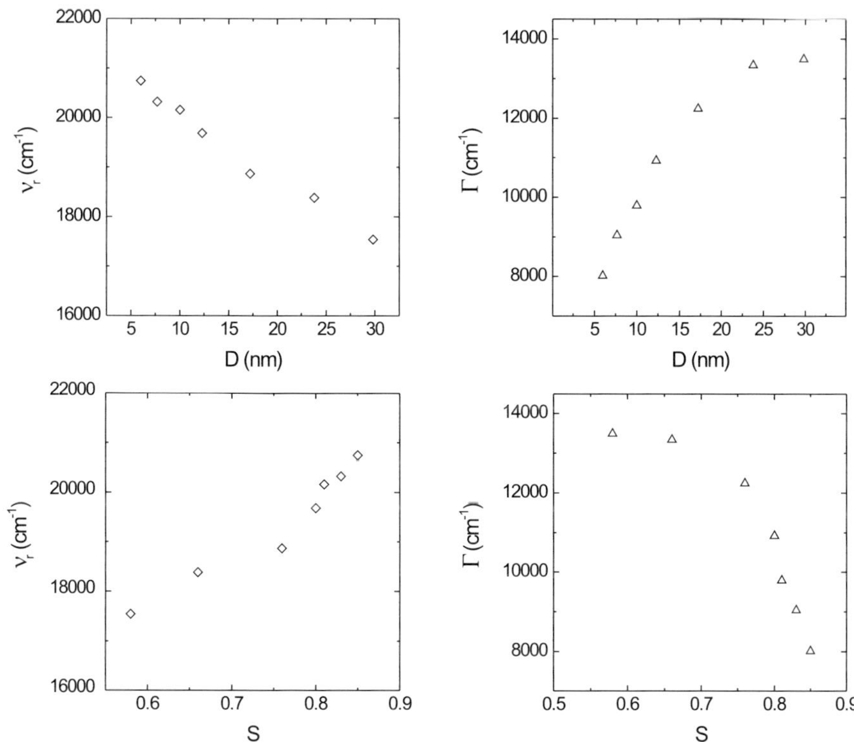

Fig. 6.28. Spectral position of the plasma resonance absorption (*left*) and mean half-width Γ (*right*) as functions of the mean particle size $\bar{D}$ (*top*) and mean shape factor $\bar{S}$ (*bottom*), respectively

Figure 6.29 shows TEM micrographs and three-dimensional particle size and shape histograms for a plasma polymer multilayer with embedded silver nanoparticles (film II/3) before and after thermal treatment up to 450 K. Experimental and calculated transmission spectra (Bergman theory) are given, as are the spectral densities $g(x)$ used in the calculations.

Although the mean particle size increases from $\bar{D} = 25.6 \pm 8.2$ nm to $\bar{D} = 31.6 \pm 14.8$ nm and the mean shape factor increases from $\bar{S} = 0.68 \pm 0.17$ to $\bar{S} = 0.72 \pm 0.13$, only the three-dimensional particle size and shape histograms demonstrate the dramatic nanostructural changes. Before thermal treatment, only a small number of particles were found with $D \geq 50$ nm and many with $S \leq 0.5$. After thermal treatment, there are more silver particles with $D \geq 50$ nm, but the number of particles with $S \leq 0.5$ has decreased dramatically. Changes in particle shape are the main reason for the spectral shift of the transmission minimum from $\tilde{\nu}_\mathrm{r} = 17\,700$ cm^{-1} to $\tilde{\nu}_\mathrm{r} = 20\,400$ cm^{-1} with $\Delta\tilde{\nu}_\mathrm{r} = 2\,700$ cm^{-1}, the decrease in half-width from $\Gamma = 15\,300$ cm^{-1} to

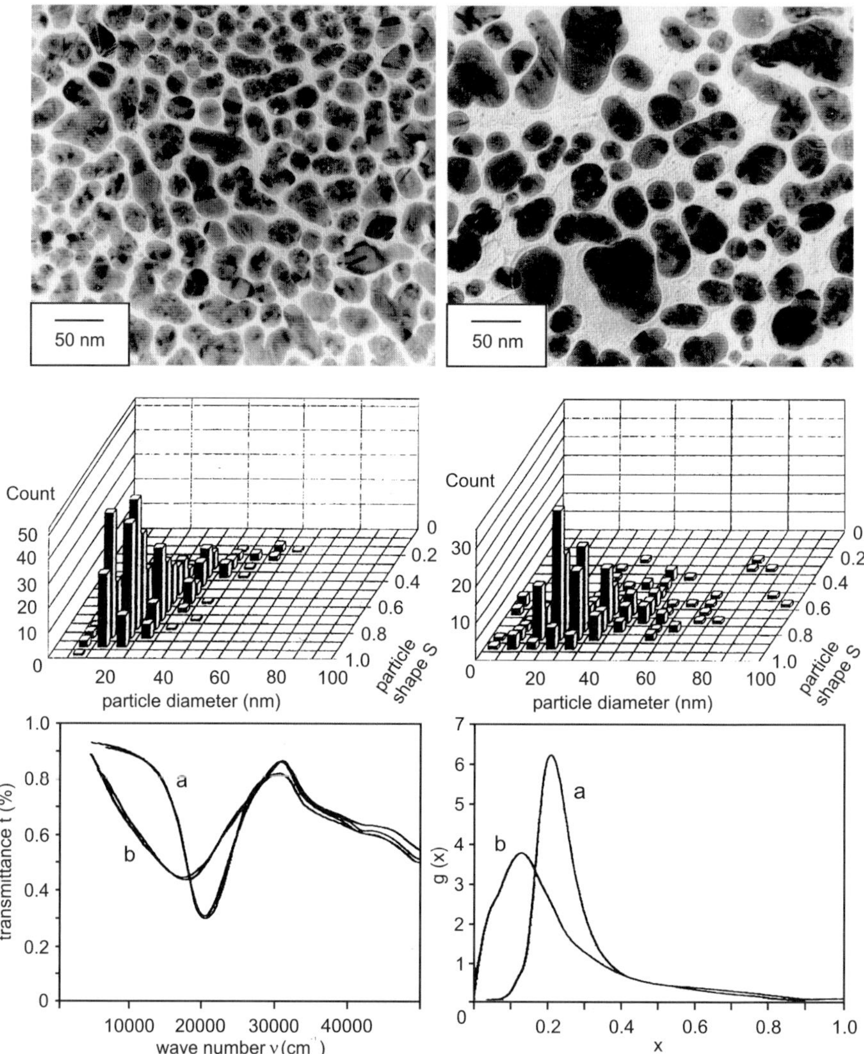

Fig. 6.29. Nanostructure and optical properties of plasma polymer films with embedded silver nanoparticles before (*left*) and after (*right*) thermal treatment up to 450 K. TEM micrographs, three-dimensional particle size and shape histograms, experimental and calculated transmission spectra (Bergman theory) (*bottom left*), and spectral densities $g(x)$ used for the calculations (*bottom right*)

$\Gamma = 7\,700 \text{ cm}^{-1}$ with $\Delta\Gamma = 7\,600 \text{ cm}^{-1}$ and the increase in relative intensity from $I_\mathrm{m} = 1.07$ to $I_\mathrm{m} = 1.84$. The changes in nanostructure can be simulated qualitatively using the Maxwell–Garnett theory and quantitatively using the Bergman theory with spectral density $g(x)$ which acts as fit function.

For the first time, optical properties of metal nanoparticles have been described with special consideration of particle size and shape distributions. It has been demonstrated that it is not sufficient to use only the mean particle diameter when describing the spectral behavior of the plasma resonance absorption for particles which deviate substantially from spherical shape. For such particle assemblies, detailed descriptions of optical properties can only be made in correlation with knowledge of the nanostructure. For theoretical descriptions, the exact and statistical routes give comparable results if an additional particle shape parameter is introduced. If the size and shape of the nanoparticles is well known, optical spectra can be calculated for each particle and summed to spectra for the whole particle assembly. This is a way to describe optical near-field experiments, for example.

References

1. M. Lee (Ed.): *International Encyclopedia of Composites*, VCH New York (1990–1994)
2. J. Delmonte: *Metal/Polymer Composites*, Van Nostrand Reinhold New York (1990)
3. H.S. Nalva (Ed.): *Nanostrucutred Materials and Nanotechnology*, Academic Press San Diego (2001)
4. T.S. Chou (Ed.) *Structure an Properties of Composites*; VCH Weinheim (1993)
5. C.F. Bohren, D.R. Huffman: *Absorption and Scattering by Small Particles*, Wiley, New York (1983)
6. U. Kreibig, M. Vollmer: *Optical Properties of Metal Clusters*, Springer Series in Material Science 25, Springer Berlin 1995
7. V.A. Madel, T.F. George (Eds.): *Optics of nanostrucutred Materials*, John Wiley & Sons New York (2001)
8. R. Kubo: J. Phys. Soc. Jpn. **17**, 975 (1962); A. Kawabata, R. Kubo: J. Phys. Soc. Japan , **21**, 1765 (1966); R. Kubo, A. Kawabata, S. Kobayashi: Ann. Rev. Mater. Sci. **14**, 49 (1984)
9. J.A.A.J. Perenboom, P. Wyder, F. Meier: Phys. Rep. **78**, 173 (1981)
10. W.P. Halperin: Rev. Mod. Phys. **58**, 533 (1986)
11. H. Haberland (Ed.): *Cluster of Atoms and Molecules* Springer (1995)
12. K.-H. Meiwes-Broer: *Metal Clusters at Surfaces*, Springer (2000)
13. G.L. Allen, R.A. Bayles, W.W. Gile, W.A. Jesser: Thin Solid Films **144**, 297 (1986)
14. B. Weitzel, H. Micklitz: Phys. Rev. Lett. **66**, 385 (1991)
15. G. Schmid: *Clusters and Colloids*, VCH Weinheim (1993)
16. J.H. Fendler: *Nanoparticles and Nanostructured Films*, WILEY-VCH (1998)
17. P. Braunstein, L.A. Oro, P.R. Raithby (Eds.): *Metal Clusters in Chemistry*, WILEY-VCH (1999)
18. F. Caruso: Adv. Mater. **13**, 11 (2001)
19. J.C. Angus, P. Koidl, S. Domitz: in *Plasma Deposited Thin Films*, J. Mort, F. Jansen (Eds.), CRC Press Boca Raton, Florida (1986)
20. D.R. McKenzie, R.C. McPhedran, N. Savvides, L.C. Botten: Phil. Mag. **48**, 341 (1983)
21. J. Robertson, Prog. Sol. State Chem. **21**, 199 (1991)
22. S.R.P. Silva (Ed.) *Amorphous carbon: state of the art*, World Scientific, Singapore (1998)
23. H. Yasuda: *Plasma Polymerization*, Academic Press, Orlando (1985)
24. R. d'Agostino (Ed.): *Plasma Deposition, Treatment, and Etching of Polymers*, Academic Press, San Diego (1990) S. 95
25. N. Inagaki: *Plasma Surface Modification and Plasma Polymerization*, Technomic Lancaster (1996)

26. H. Yasuda: J. Polym. Sci.: Macromol. Rev. **16**, 199 (1981)
27. H. Biederman, D. Slavinska: Surf. Coat. Technol. **125**, 371 (2000)
28. H.U. Poll: PhD Thesis B, TH Karl-Marx-Stadt (1975)
29. E. Kay, J. Coburn, A. Dilks: Topics in Current Chemistry **94**, S. Veprek, M. Venugopalan (Eds.); Springer, Berlin (1980) P. 1
30. A.T. Bell: Topics in Current Chemistry **94**, S. Veprek, M. Venugopalan (Eds.), Springer, Berlin (1980) P. 44
31. S. Kanazawa, M. Kogoma, T. Moriwaki, S. Okazaki: J. Phys. D: Appl. Phys. **21**, 838 (1988); T. Yokohama, M. Kogoma, S. Kanazawa, T. Moriwaki: J. Phys. D: Appl. Phys. **23**, 374 (1990)
32. N. Morosoff: in *Plasma Deposition, Treatment, and Etching of Polymers*, R. d'Agostino (Ed.), Academic Press, San Diego, (1990) P. 1
33. H.U. Poll, M. Arzt, K.H. Wickleder: Europ. Polym. J. **12**, 505 (1976)
34. P. Favia, F. Fracassi, R. d'Agostino: J. Biomater. Sci. Polymer Edn. **4**, 61 (1992)
35. S. Sahli, Y. Segui, S. Hadj Moussa, M.A. Djouadi: Thin Solid Films **217**, 17 (1992)
36. R. d'Agostino, F. Cramarossa, F. Fracassi, F. Illuzi: in *Plasma Deposition, Treatment, and Etching of Polymers*, R. d'Agostino (Ed.), Academic Press, San Diego (1990) P. 95
37. R. d'Agostino, F. Fracassi, F. Illuzi: Polym. Mat. Sci. Engn. **62**, 157 (1991)
38. R.K. Sadhir, K.F. Schoch Jr.: Thin Solid Films **223**, 154 (1993)
39. A. Kruse, A. Baalmann, W. Budden, V. Schlett, M. Hennecke: Surf. Coat. Technol. **59**, 359 (1993)
40. A. Kiesow, A. Heilmann: Thin Solid Films **343–344**, 338 (1999)
41. A.M. Wrobel, M.R. Wertheimer: in *Plasma Deposition, Treatment, and Etching of Polymers*, R. d'Agostino (Ed.), Academic Press, San Diego (1990) P. 163
42. K. Matsui, S. Yoshida: J. Appl. Phys. **64**, 2607 (1988)
43. I. Terada, T. Haraguchi, T. Kajiyama: Polym. J. **18**, 529 (1986)
44. A.M. Wrobel: Plasm. Chem. Plasm. Proc. **7**, 429 (1987)
45. A. Kiesow, A. Heilmann, to be published
46. R.L. Crawley, J.L. Evans: J. Vac. Sci. Technol. A **9**, 824 (1991)
47. T. Ihara, H.K. Yasuda: Polym. Mat.: Sci. Engn. **62**, 32 (1990)
48. J. Kieser, M. Sellschopp: J. Vac. Sci. Technol. A **4**, 222 (1986)
49. T.J. Lin, B.H. Chun, H.K. Yasuda, D.J. Yang, J.A. Antonelli: J. Adhesion Sci. Technol. **5**, 893 (1991)
50. Y. Lin, H. Yasuda: J. Appl. Polym. Sci. **60**, 543 (1996)
51. M.R. Wertheimer, J.E. Klemberg-Sapieha, H.P. Schreiber: Thin Solid Films **115**, 109 (1984)
52. R.K. Sadhir, H.E. Saunders: Semiconduct. Int. **3**, 110 (1985)
53. G. Akovali: J. Appl. Polym. Sci. **32**, 4027 (1986)
54. T. Wydeven, C.C. Johnson: Polym. Prep. Am. Chem. Soc., Div. Polym. Chem. **21**, 62 (1982)
55. C. Oehr, B. Schindler: Surf. Coat. Techn. **98**, 848–850 (1998)
56. J. Xinyu, Z. Yu, W. Rui, C. Kangsheng, L. Suzhen: J. Vac. Sci. Technol B **15**, 2688 (1997)
57. L. Mex, A. Kramer, C. Francke, J. Müller: Supercond. Sci. Techn. **11**, 426 (1998)
58. C.E. Moffitt, C.M. Reddy, Q.S. Yu, D.M. Wieliczka, H.K. Yasuda, Appl. Surf. Sci. **161**, 481 (2000)
59. B. Schumacher, W. Müschenborn, M. Stratmann, B. Schultrich, C.-P. Klages, M. Kretschmer, U. Seyfert, F. Förster, H.-J. Tiller: Adv. Engin. Mater. **3**, 681 (2001)

60. I. Retzko, J.F. Friedrich, A. Lippitz, W.E.S. Unger: J. Electr. Spectros. Rel. Phenomena **121**, 111 (2001)
61. H.U. Poll, S. Schreiter: Surface and Coating Technology **93**, 105 (1997)
62. J. Weichart, J. Müller, S. Nehlsen: 10th Int. Symp. Plasma Chemistry Bochum 1991, Symp. Proc., U. Ehlemann, H.G. Lergon, K. Wiesemann (Eds.), **Band 2**, S. 2, 3–5 (1991)
63. N. Inagaki, S. Kondo, M. Hirata, H. Urushibata: J. Appl. Polym. Sci. **30**, 3385 (1985)
64. N. Inagaki, H. Katsuoka: J. Membrane Sci. **34**, 297 (1987); N. Inagaki, N. Kobayashi, M. Matsushima: J. Membrane Sci. **38**, 85 (1988)
65. H. Nomura, P.W. Kramer, H. Yasuda: Thin Solid Films **118**, 187 (1984)
66. H. Matsuyama, K. Hirai, M. Teramoto: J. Membrane Science **92**, 257 (1994)
67. J. Weichart, J. Müller: Surf. Coat. Technol. **59**, 342 (1993)
68. T. Kashiwagi, K. Okabe, K. Okita: J. Membrane Sci. **36**, 353 (1988)
69. Y. Uchimoto, K. Yasuda, Z. Ogumi, Z. Takehara: J. Electrochem. Soc. **138**, 3190 (1991)
70. S. Nehlsen, T. Hunte, J. Müller: J. Membrane Sci. **1**, 106 (1995)
71. M. Walker, K.M. Baumgärtner, J. Feichtinger, M. Kaiser, E. Räuchle, J. Kerres: Surf. Coat. Technol. **116–119**, 996 (1999)
72. K. Li, J. Meichsner: Surf. Coat. Techn. **116–119**, 841–847 (1999)
73. A.W. Hahn, D.H. York, M.F. Nichols, G.C. Amormin, H.K. Yasuda: J. Appl. Polym. Sci.: Appl. Polym. Symp. **38**, 55 (1984)
74. H.U. Poll: Z. exp. Chir. Transplant. Künstl. Organe **17**, 306 (1984)
75. B.D. Ratner, A. Chilkoti, G.P. Lopez: in *Plasma Deposition, Treatment, and Etching of Polymers*, R. d'Agostino (Ed.), Academic Press San Diego (1990) P. 463
76. G. Kampfrath, R. Hintsche: Anal. Lett. **22**, 2423 (1989)
77. Y. Jimbo, M. Saito: J. Molec. Electr. **4**, 111 (1988)
78. D. Janasek, U. Spohn, A. Kiesow, A. Heilmann: Sens. Actuat. B **78**, 228 (2001)
79. C. Hamann, G. Kampfrath, M. Müller: Sensors Actuators B **1**, 142 (1990)
80. S. Takeda: Vacuum **41**, 1769 (1990)
81. N. Inagaki, K. Suzuki, K. Nejigaki: J. Appl. Polym. Sci.: Polym. Lett. Ed. **21**, 353 (1983); N. Inagaki: Thin Solid Films **118**, 225 (1984)
82. E. Radeva, K. Bobev, L. Spassov, I. Tzolovski: Dokl. Bulg. Akad. Nauk **44**, 39 (1991)
83. E. Radeva, Vacuum bf 48, 41 (1997); E. Radeva, L. Spassov: Vacuum **52**, 217 (1998)
84. A.R.K. Ralston, J.A. Tobin, S.S.Bajikar, D.D Denton: Sens. Actuat. B **22**, 139 (1994)
85. I. Sugimoto, M. Nakamura, H. Kuwano: Sensors and Actuators B **10**, 117 (1993)
86. N. Inagaki, J. Ohkubo: J. Appl. Polym. Sci. **43**, 793 (1991)
87. J. Janca, L. Sodomka: Surf. Coat. Techn. **98**, 851–854 (1998)
88. Y. Matsuda, H. Yasuda: Thin Solid Films **118**, 211 (1984)
89. E. Dayss, G. Leps, J. Meinhardt: Surf. Coat. Technol. **116–199**, 986 (1998)
90. H. Yamada, M. Nakamura, S. Morita, S. Hattori: SPIE Proc. **1086**, 312 (1986)
91. M.W. Horn, S.W. Pang, M. Rothschild: J. Vac. Sci. Technol. B **8**, 1493 (1990)
92. V.S. Nguyen, J. Underhill, S. Fridmann, P. Pan: J. Electrochem. Soc. **132**, 1925 (1985)
93. Y. Kagami, K. Yamada, T. Yamauchi, J. Gong, Y. Osada: J. Appl. Polym. Sci.: Appl. Polym. Symp. **46**, 289 (1990)
94. S. Morita, J. Tamano, S. Hattori, M. Ieda: J. Appl. Phys. **51**, 3938 (1980)
95. S. Morita, S. Hattori: Pure Appl. Chem. **57**, 1277 (1985)

96. S.A. Gangal, M. Hori, S. Morita, S. Hattori: Thin Solid Films **149**, 341 (1987)
97. F.O. Fong, K.C. Kuo, J.C. Wolfe, J.N. Randall: J. Vac. Sci. Technol. B **6**, 375 (1988)
98. M. Hori, S. Hattori, T. Yoneda, S. Morita: J. Vac. Sci. Technol. B **4**, 500 (1986)
99. H. Hirai, H. Sekiguchi, S. Miyata, S. Kobayashi: Appl. Phys. Lett. **50**, 818 (1987)
100. V.J. Rao, V. Manorama, S.V. Bhoraskar: Appl. Phys. Lett. **54**, 1799 (1989); M. Vardhireddy, R.S. Bhide, S.V. Bhoraskar, V.J. Rao: Polym. Mat.: Sci. Engn. **62**, 373 (1990)
101. Z.J. Horvath: J. Appl. Phys. **68**, 5901 (1990)
102. Y. Segui, D. Montalan, B. Moret: Thin Solid Films **120**, 37 (1984)
103. C. Hamann, G. Kampfrath: Vacuum **34**, 1053 (1984)
104. J. Leiber, W. Michaeli: ANTEC 92 Conf. Detroit, Conf. Proc. S. 2583 (1992)
105. R.K. Sadhir, H.E. Saunders: Proc. 17th IEEE Electr. Electron. Conf. Boston (1985) S. 282
106. D.H. Davies, R.H. Miller, S. Maki, D. Devine: Polym. Mat.: Sci. Engn. **62**, 343 (1990)
107. K. Nagawa: J. Appl. Polym. Sci. 41, 2049 (1990).
108. Y. Koike, E. Nihei, N. Tanio, Y. Ohutsuka: Appl. Opt. 29, 2686 (1990).
109. A.R. Nyaiesh, L. Holland: Vacuum **34**, 519 (1984)
110. M. Kusabiraki: J. Appl. Polym. Sci.: Appl. Polym. Symp. **46**, 473 (1990)
111. J. Vlassak, W. Nix, A. Heilmann, E. Kay: unpublished results
112. K.W. Gerstenberg: Coll. Polym. Sci. **268**, 345 (1990)
113. A. Kiesow, H. Knoll, M. Petzold, A. Heilmann, unpublished results
114. S.Y. Park, N. Kim: Polym. Mat.: Sci. Engn. **62**, 152 (1990); S.Y. Park, N. Kim: J. Appl. Polymer Science: Appl. Polym. Symp. **46**, 91 (1990)
115. B. Dischler, A. Bubenzer, P. Koidl: Appl. Phys. Lett. **42**, 636 (1983)
116. A.M. Wrobel, M.R. Wertheimer, J. Dib, H.P. Schreiber: J. Macromol. Sci. Chem. A **14**, 321 (1980)
117. A. Morinaka, Y. Asano: J. Appl. Polym. Sci. **27**, 2139 (1982)
118. H.U. Poll, J. Meichsner, M. Arzt, M. Friedrich, R. Rochotzki, E. Kreyàig: Surf. Coat. Technol. **59**, 365 (1993)
119. C. Rau, W. Kulisch: Thin Solid Films **249**, 28 (1994)
120. S. Sahli, Y. Segui, S. Ramdani, Z. Takkouk: Thin Solid Films **250**, 206 (1994)
121. L. Zuri, M.S. Silverstein, M. Narkis: J. Appl. Polym. Sci. **62**, 2147 (1996)
122. S. Eufinger, W.J. cvan Ooij, K.D. Connors: Surf. Interf. Anal. **24**, 841 (1996)
123. A.M. Wrobel: J. Macromol. Sci. Chem. A **22**, 1089 (1985)
124. J. Tyczkowski, H. Szymanowski: Mat. Sci. Engn. A **139**, 120 (1991)
125. T. Danzer, G. Marx, G. Riedel: Thin Solid Films **219**, 119 (1992)
126. F.M. Tibbit, M. Shen, A.T. Bell: J. Macromol. Sci. Chem. A **10**, 1623 (1976)
127. Bulletin 137–78, Commonwealth Scientific Co. Alexandria VI (1978)
128. R.J. Buss: J. Appl. Phys. **59**, 2977 (1986)
129. S. Schelz, P. Reinke, P. Oelhafen: Surf. Sci. **293**, 275 (1993)
130. M.J. Vasile, G. Smolinsky: J. Electrochem. Soc. **119**, 451 (1972)
131. H.T. Chiu, J.S. Hsu: Thin Solid Films **252**, 13 (1994)
132. G. Li, J.A. Tobbin, D.D. Denton, Appl. Phys. Lett. **64**, 560 (1994)
133. M. Yamana, K. Kanda, N. Kashiwazaki, M. Yamamoto, T. Nakano, C. Walton: Jap. J. Appl. Phys. **L28**, 1592 (1989)
134. T. Tchoubineh, S. Mondin, F. Arefi, J. Amouroux: Polym. Mat.: Sci. Engn. **62**, 548 (1990)
135. T.S. Ramu, M.R. Wertheimern, F.R. Klemberg-Sapieha: IEEE Trans. Electr. Insulation **EI 21**, 549 (1986)

136. W. Müller, M. Schmidt, R. Seefeldt, R. Wilberg: Beitr. Plasmaphys. **24**, 629 (1984)
137. L.F. Thompson, G. Smolinsky: J. Appl. Polym. Sci. **16**, 1179 (1972)
138. D. Hafer, H.J. Tiller, K. Meyer: Plaste und Kautschuk **19**, 354 (1972)
139. G.W. Collins, S.A. Letts, E.M. Fearon, R.L. McEachern, T.P. Bernat: Phys. Rev. Lett. **73**, 708 (1994)
140. W. Michaeli, K. Telgenbüscher, J. Striffler,J. Leiber, M. Stollenwerk: Surf. Coat. Technol. **59**, 338 (1993)
141. H. Haken, H.C. Wolff: *Molekülphysik und Quantenchemie*, Springer, Berlin (1994)
142. S. Onari: J. Phys. Soc. Japan **260**, 500 (1969)
143. T. Wydeven, P. Wood, O. Tsuji: Polym. Mat.: Sci. Engn. **62**, 338 (1990)
144. R.P. Mota, D. Galvao, S.F. Durrant, M.A. Bica de Moraes, S.O. Dantas, M. Cantao: Thin Solid Films **270**, 109 (1995)
145. J. Tauc, R. Grigorovici, A. Vancu: Phys. Stat. Sol. **15**, 627 (1966)
146. R. Petrich: PhD Thesis, TU Chemnitz-Zwickau (1995)
147. S.F. Durrant, R.P. Mota, M.A. Bica de Moraes: J. Appl. Phys. **71**, 448 (1992)
148. H. Biedermann, L. Martinu, J. Zemek: Vacuum **35**, 447 (1985)
149. G. Robertson: Adv. Phys. **35**, 317 (1986)
150. A. Bradley, J.P. Hammes: J. Electrochem. Soc. **110**, 15 (1963)
151. D. Brosset, B. Al, Y. Segui: Appl. Phys. Lett. **33**, 87 (1978)
152. S. Sapieha, M. Kryszwski: Arch. Elektrotech. (Warszawa) **23**, 739 (1974)
153. D. Montalan, N. Souag, Y. Segui, C. Laurent: Thin Solid Films **170**, 235 (1989)
154. K. Ishii, Y. Ohki, T. Nakano: Conf. Rep. IEEE Int. Symp. Electr. Insulation Toronto (1990) S. 76
155. K. Tanaka, K. Yoshizawa, T. Takeuchi, T. Yamabe:Synthetic Metals **38**, 107 (1990)
156. K. Tanaka, T. Yamabe, T. Takeuchi, K. Yoshizawa, S. Nishio: J. Appl. Phys. **70**, 5653 (1991)
157. H. Grünwald, H.S. Munro, T. Wilhelm: Mat. Sci. Engn. A **139**, 356 (1991)
158. K. Tanaka, S. Nishio, Y. Matsura, T. Yamabe: J. Appl. Phys. **73**, 5017 (1993)
159. Y. Osada, Q.S. Yu, H. Yasunaga, F.S. Wang, J. Chen: J. Appl. Phys. **64**, 1476 (1988)
160. X. Xie, J.U. Thiele, R. Steiner, P. Oelhafen: Synthetic Metals **63**, 221 (1995)
161. Y. Takai, M. Inoue, T. Mizutani, M. Ieda: J. Phys. D.: Appl. Phys. **18**, 1637 (1985)
162. G. Sawa, T. Murata, M. Ohara, F. Sunda, S. Nakamura: Jap. L. Appl. Phys. **25**, 53 (1986)
163. Z. Ogumi, Y. Uchimoto, Z. Takehara: Polym. Mat.: Sci. Engn. **62**, 363 (1990). Y. Uchimoto, Z. Ogumi, Z. Takehara: Solid State Ionics **40/41**, 624 (1990); J. Electrochem. Soc. **137**, 35 (1990)
164. A.M. Okoniewski, A. Yelon: J. Appl. Phys. **58**, 414 (1985); J. Tyczkowski, G. Czeremuszkin, M. Kryszweski: Phys. Stat. Sol (a) **72**, 751 (1982)
165. M. Gazicki, H. Yasuda: Plasma Chem. Plasma Proc. **3**, 179 (1983)
166. P.K. Abraham, K. Sathianandan: Thin Solid Films **164**, 353 (1988)
167. J. Tyczkowski: Thin Solids Films **199**, 335 (1991)
168. G. Schmid: Adv. Enginoor. Mat. **3**, 737 (2001)
169. H.S. Göktürk, T.J. Fiske, D.M. Kalyon: J. Appl. Phys. **73**, 5598 (1993)
170. R. Pelster, G. Galeczki, G. Nimtz, P. Pissis: J. Non. Crystall. Solids **131–133**, 238 (1991)
171. S.I. Lee, T.W. Noh, J.R. Gaines: Phys. Rev. B **32**, 3580 (1985) T.W. Noh, S.I. Lee, J.R. Gaines: Phys. Rev. B **33**, 1401 (1985); T.W. Noh, S.I. Lee, J.R. Gaines: Phys. Rev. B **34**, 3882 (1985)

172. D. Fornasiero, F. Grieser: Chem. Phys. Lett. **139**, 103 (1987)
173. N. Satoh, H. Hasegawa, K. Tsujii, K. Kimura: J. Phys. Chem. **98**, 2143 (1994)
174. J.G. Skofronick, W.B. Phillips: J. Appl. Phys. **38**, 4791 (1967)
175. F. Wu, J.E. Morris: Thin Solid Films **246**, 17 (1994); J.E. Morris, A. Kiesow, M. Hong, F. Wu: Int. J. Electron. **81**, 441 (1996)
176. N.D. Uan: Thesis A, TU Karl-Marx-Stadt (1988)
177. T. Noguchi, K. Gotoh, Y. Yamaguchi, S. Deki: J. Mater. Sci. Lett. **10**, 477 (1991); **10**, 648 (1991); **11**, 797 (1992)
178. K. Akamatsu, S. Deki: J. Mater. Chem. **7**, 1773 (1997); **8**, 673 (1998); Scripa mater. **44**, 2149 (2001)
179. K. Akamatsu, S. Deki: Nanostruc. Mat. **8**, 1121 (1998); K. Akamatsu, M. Tsuboi, Y. Hatakenaka, S. Deki: J. Phys. Chem. B **104**, 10168 (2000)
180. S. Deki, K. Akamatsu, Y. Hatakenaka, M. Mizuhata, A. Kajinami: Nanostruc. Mat., **11**, 59 (1999)
181. Y. Inoue, M. Fujii, M. Inata, S. Hayashi, K. Yamamoto, K. Akamatsu, S. Deki: Thin Solid Films,**372**, 169 (2000)
182. K. Akamatsu, S. Takei, M. Mizuhata, A. Kajinami, S. Deki, S. Takeoka, M. Fujii, S. Hayashi, K. Yamamotot: Thin Solid Films **359**, 55 (2000)
183. M. Fujii, Y. Inoue, S. Hayashi, K. Akamatsu, S. Deki: Jpn. J. Appl. Phys. **40**, 1911 (2001)
184. S. Deki, H. Nabika, K. Akamatsu, M. Mizuhata, A. Kajinami: Scipta mater. **44**, 1879 (2001)
185. A. Nannini, P.E. Bagnoli, A. Diligenti, B. Neri, S. Pugliese: J. Appl. Phys. **62**, 2138 (1987)
186. R.A. Roy, R. Messier, V. Krishnaswamy: Thin Solid Films **109**, 27 (1983)
187. A. Kiesow, G. Willers, N. Teuscher, A. Heilmann: in preparation
188. A. Nannini, F. Cacialli, P. Bruschi: J. Mol. Electr. **7**, 21 (1991) P. Bruschi, A. Nannini, F. Massara: Thin Solid Films **196**, 201 (1991)
189. D. Babonneau, T. Cabioc'h, A. Naudon, J.C. Girard, M.F. Denanot: Surf. Sci. **409**, 358 (1998)
190. Y. Inoue, M. Inata, M. Fujii, S. Hayashi, K. Yamamoto: Thin Solid Films **349**, 289 (1999)
191. Y. Inoue, T. Kita, S. Hayashi, K. Yamamoto: J. Phys.: Condes. Matter **9**, 8669 (1997)
192. M. Fujii, T. Nagareda, S. Hayashi, K. Yamamoto: Phys. Rev. B **44**, 6243 (1991)
193. R.W. Cohen, G.D. Cody, M.D. Coutts, B. Abeles: Phys. Rev. B **8**, 3689 (1973)
194. M. Gadenne, P. Gadenne, J.C. Martin, C. Sella: Thin Solid Films **221**, 183 (1992)
195. B. Abeles, Ping Sheng, M.D. Coutts, Y. Arie: Adv. Phys. **24**, 407 (1975)
196. T.E. Schlesinger, R.C. Cammarata, A. Gavrin, J.Q. Xiao, C.L. Chien, M.K. Ferber, C. Hayzelden: J. Appl. Phys. **70**, 3275 (1991)
197. J.I. Gittleman, B. Abeles, P. Zanzucchi, Y. Arie: Thin Solid Films **45**, 9 (1977)
198. L.K. Thomas, C. Tang: Solar Energy Mat. **18**, 117 (1989)
199. N. Tanaka, K. Mihama: J. Cryst. Growth **99**, 577 (1990)
200. K. Mihama, S. Iwama, N. Tanaka: Z. Phys. D **26**, S243 (1993)
201. Y. Inoue, M. Fujii, S. Hayashi, K. Yamamoto: Solid-State Electr. **42**, 1605 (1998)
202. E.M. Logothetis, W.J. Kaiser, H.K. Plummer, S.S. Shinozaki: J. Appl. Phys. **60**, 2548 (1986); W.J. Kaiser, E.M. Logothetis, L.E. Wenger: Sol. State Comm. **58**, 83 (1986)
203. S. Hayashi, M. Fujii, K. Yamamoto: Jap. J. Appl. Phys. **28**, L1464 (1989)

204. M. Fujii, S. Hayashi, K. Yamamoto: Appl. Phys. Lett. **57**, 2692 (1990); M. Fujii, S. Hayashi, K. Yamamoto: Jap. J. Appl. Phys. **L 30**, 687 (1991); M. Fujii, M. Wada, S. Hayashi, K. Yamamoto: Phys. Rev. B **46**, 15930 (1992)
205. S. Hayashi, T. Nagareda, Y. Kanazawa, K. Yamamoto: Jap. J. Appl. Phys. A **32**, 3840 (1993)
206. S. Berthier, J. Lafait, C. Sella, T.K. Vien: Thin Solid films **125**, 171 (1985)
207. T.E. Schlesinger, A. Gavrin, R.C. Cammarata, C.L. Chien: Mat. Res. Soc. Symp. Proc. **195**, 441 (1990)
208. A. Tsoukatos, G.C. Hadjipanayis: J. Appl. Phys. **70**, 5891 (1991)
209. A.S. Edelstein, B.N. Das, R.L. Holtz, N.C. Koon, M. Rubinstein, S.A. Wolf, K.E. Kihlstrom: J. Appl. Phys. **61**, 3320 (1987)
210. S.H. Liou, C.L. Chien: Appl. Phys. Lett. **52**, 512 (1988)
211. J.L. Dormann, D. Fiorani: Mat. Res. Soc. Symp. Proc. **195**, 429 (1990)
212. A. Gavrin, C.L. Chien: J. Appl. Phys. **67**, 938 (1990)
213. N. Boonthanom, M. White: Thin Solid Films **24**, 295 (1974)
214. S.S. Elovikov, V.P. Novojilov, N.V. Ushakov: Thin Solid Films **62**, 303 (1979)
215. A. Shibukawa: Appl. Opt. **18**, 1460 (1979)
216. U. Kreibig, D. Andersson, G.A. Niklasson, C.G. Granqvist: Thin Solid Films **125**, 199 (1985)
217. W. Pamler, E.E. Marinero, M. Chen: Phys. Rev. B **33**, 5736 (1986)
218. K. Sujatha, S. Uthanna, P.J. Reddy: J. Mater. Sci. Lett. **8**, 75 (1989)
219. J.F. Roux, M. Treilleux, B. Cabaud, C. Montandon, C. Martet, A. Hoareau: Conf. Proc. ICEM **13** (Paris 1994) 376; J.F. Roux, M. Treilleux, B. Cabaud, G. Fuchs, C. Montandon, A. Hoareau: J. Phys I (Paris) **4**, 991 (1994)
220. G. Boivin, J.M. Theriault: Appl. Opt. **23**, 4245 (1984)
221. G.A. Niklasson, C.G. Granqvist: J. Appl. Phys. **55**, 3382 (1984)
222. M.F. MacMillan, R.P. Devaty, J.V. Mantese: Mat. Res. Symp. Proc. **132**, 99 (1989)
223. E. Kny, L.L. Levenson, W.J. James, R.A. Averbach, J. Vac. Sci. Technol. **16**, 359 (1979); J. Phys. Chem. **84**, 1635 (1980)
224. R.K. Sadhir, H.E. Saunders, W.J. James: ACS Symp. Ser. **242**, 533 (1984)
225. N.C. Morosoff, C.D. Clymer, J.M. Skelly, A.L. Crumbliss: J. Appl. Polym. Sci.: Appl. Polym. Symp. **46**, 315 (1990)
226. M. Gazicki, Chaos, Solitons and Fractals **10**, 1983 (1999)
227. M. Gazicki, W. Fallmann, A. Jachimowicz, F. Kohl, F. Olcaytug, K. Pirker, R. Schallauer, G. Urban: Polym. Mat.: Sci. Engn. **62**, 368 (1990); M. Gazicki, H. Szymanowski, J. Schalko, L. Malinovsky: Thin Solid Films **230**, 81 (1993); M. Gazicki, H. Szymanowski, J. Tyczkowski, J. Schalko, F. Olcaytug: J. Vac. Sci. Technol. A **14**, 2835 (1996); M. Gazick, R. Ledzion, R. Madzurczyk, S. Pawlowski: Thin Solid Films **322**, 123 (1998)
228. R. Brusasco, M. Saculla, R. Cook: J. Vac. Sci. Technol. A **13**, 948 (1995)
229. R.K. Sadhir, W.J. James: ACS Symp. Ser. **242**, 555 (1984)
230. C. Oehr, H. Suhr: Thin Solid Films **155**, 65 (1987)
231. J. Kuhn, S. Lasch, H. Lehmberg, H. Pagnia: Int. J. Electron. **73**, 1003 (1992)
232. Y. Kagami, T. Yamguchi, Y. Osada: J. Appl. Phys. **68**, 610 (1990)
233. N. Morosoff, D.L. Patel, A.R. Write, M. Umana, A.L. Crumbliss, P.S. Lugg, D.B. Brown: Thin Solid Films **117**, 33 (1984)
234. N. Morosoff, R. Haque, S.D. Clymer, A.L. Crumbliss: J. Vac. Sci. Technol. A **3**, 2098 (1985)
235. M. Yamada, J. Tamano, K. Yoneda, S. Morita, S. Hattori: Jap. J. Appl. Phys. **21**, 768 (1982)
236. H. Suhr, A. Etspüler, E. Feurer, C. Oehr: Plasma Chem. Plasma Proc. **8**, 9 (1988)

237. F. Homilius, A. Heilmann, C. von Borczyskowski: Surf. Coat. Technol. **74–75**, 594 (1995); F. Homilius, A. Heilmann, U. Rempel, C. von Borczyskowski: Vacuum **49**, 205 (1998)
238. J.W. Coburn, E. Kay: Plasma Method for Forming a Metal Containing Polymer, US Pat. **4**, 226, 896 (1980)
239. C. Hamann, A. Heilmann, D.-E. Endres: Verfahren zur Herstellung von punktförmigen Quanten-Bauelementen DE 4200193 C2 (1992)
240. B. Despax, E. Kay: Circuit Writing Process, US Pat. **4**, 794, 087 (1988)
241. P.B. Comita, E. Kay, R. Zhang, W. Jacob: Appl. Surf. Sci. **79/80**, 196 (1994)
242. P.B. Comita, W. Jacob, E. Kay, R. Zhang: SPIE Proc. **1804**, 154 (1993)
243. M. Kaempfe, H. Graener, G. Seifert, J. Lange, A. Kiesow, A. Heilmann: Pat. DE 100 64 456.2 (2000)
244. M. Kaempfe, H. Graener, A. Kiesow, A. Heilmann: Appl. Phys. Lett. **79**, 1876 (2001)
245. M. Hori, T. Yoneda, H. Yamda, S. Morita, S. Hattori: Plasma Chem. Plasma Proc. **7**, 155 (1987)
246. P. Canet, C. Laurent, J. Akinnifesi, B. Despax: J. Appl. Phys. **73**, 384 (1993)
247. C. Laurent, E. Kay, N. Souag: J. Appl. Phys. **64**, 336 (1988)
248. T. Elallam, J. Akinnifesi, B. Despax, T.G. Hoang, M. Saidi, M. Bendaoud: J. Appl. Phys. **76**, 3869 (1994)
249. A. Kiesow: PhD Thesis, TU Chemnitz (2001)
250. H. Yasuda, T.J.O. Keefe, D.L. Cho, B.K. Sun: Polym. Mat.: Sci. Engn. **62**, 524 (1990)
251. N. Tegen, J. Wartusch, K.-H. Merkel: Nucl. Instrum. Meth. Phys. Res. B **80/81**, 1055 (1993)
252. A. Dilks and E. Kay: ACS Symp. Ser. **108**, 195 (1979)
253. P. Canet, C. Laurent, J. Akinnifesi, B. Despax: J. Appl. Phys. **72**, 2423 (1992)
254. R. d'Agostino, L. Martinu, V. Pische: Plasma Chem. Plasma Proc. **11**, 1 (1991)
255. M. Wang, K. Schmidt, K. Reichelt, X. Jiang, H. Huebsch, H. Dimigen: J. Mater. Res. **7**, 1465 (1992)
256. E. Kay: Z. Phys. D **3**, 251 (1986)
257. E. Kay, M. Hecq: J. Appl. Phys. **55**, 370 (1984)
258. J. Perrin, B. Despax, E. Kay: Phys. Rev. B **32**, 719 (1985)
259. J. Perrin, B. Despax, V. Hanchett, E. Kay: J. Vac. Sci. Technol. A **4**, 46 (1986)
260. E. Kay, A. Dilks, U. Hetzler: J. Macromol. Sci. Chem. A **12**, 1393 (1978)
261. E. Kay, A. Dilks, D. Seybold: J. Appl. Phys. **51**, 5678 (1980)
262. E. Kay, A. Dilks: Thin Solid Films 78, 309 (1981)
263. E. Kay, M. Hecq: J. Vac. Sci. Technol. A **2**, 401 (1984)
264. M. Hecq, P. Zieman, E. Kay: J. Vac. Sci. Technol. A **1**, 364 (1983)
265. A. Heilmann, J. Werner, M. Kelly, B. Holloway, E. Kay Appl. Surf. Sci. **115**, 365–376 (1997)
266. C. Laurent, E. Kay: Z. Phys. D **12**, 465 (1989)
267. W.R. Creasy, J.A. Zimmerman, W. Jacob, E. Kay: J. Appl. Phys. **72**, 2462 (1992)
268. D. Dalcu, L. Martinu: J. Vac. Sci. Technol. A **17**, 877 (1999); D. Dalacu, A.P. Brown, J.E. Klemberg-Spieha, L. Martinu, M.R. Wertheimer, S.I. Najafi, M.A. Andrews: MRS Symp. Proc. **544**, 167–172 (1999)
269. H. Biedermann, L. Holland: Nucl. Instr. Meth. **212**, 497 (1983)
270. H. Biederman: Vacuum **34**, 405 (1984)
271. L. Martinu, H. Biederman: Vacuum **36**, 477 (1986)
272. L. Martinu, H. Biederman: Plasma Chem. Plasma Proc. **5**, 81 (1985)
273. A. Fejfar, L. Martinu, I. Ostadal: Vacuum **39**, 19 (1989)

274. L. Martinu: Thin Solid Films **140**, 307 (1986)
275. L. Martinu: Solar Energy Mater. **15**, 135 (1987)
276. L. Martinu: Solar Energy Mater. **15**, 21 (1987)
277. L. Martinu, H. Biederman: J. Vac. Sci. Technol. A **3**, 2639 (1985)
278. H. Biederman, L. Martinu, D. Slavinska, I. Chudacek: Pure Appl. Chem. **60**, 607 (1988)
279. C. Laurent, E. Kay: J. Appl. Phys. **65**, 1717 (1989)
280. J.L. Flouttard, J. Akinnifesi, E. Cambril, B. Despax: J. Appl. Phys. **70**, 798 (1991)
281. B. Despax, J.L. Flouttard: Thin Solid Films **168**, 81 (1989)
282. C. Laurent, D. Mauri, E. Kay, S.S.P. Parkin: J. Appl. Phys. **65**, 2017 (1989)
283. E. Kay, C. Laurent, S.S.P. Parkin, D. Mauri: Mat. Res. Soc. Symp. Proc. **195**, 423 (1990)
284. E. Cambril, J. Akinnifesi, J.M. Enjalbert, B. Despax: J. Phys. D. Appl. Phys. **26**, 149 (1993)
285. C. Guyard, B. Despax: Surf. Coat. Techn. **116–119**, 638–642 (1999)
286. M. Zeuner, H. Neumann, J. Zalman, D. Slavinska, H. Biederman: Vacuum **51** 417 (1998)
287. H. Biederman, P. Hlidek, J. Pesicka, D. Slavinska, V. Studzia, J. Zemek, R.J. Kingdon, R.J. Howson: Vacuum **47**, 1453 (1996); H. Biederman, R.J. Howson, D. Slavinska, V. Studzia, J. Zemek: Vacuum **48**, 883 (1997)
288. H. Biederman, V. Studzia, D. Slavinska, J. Zalman, J. Pesicka, M. Vanecek, J. Zemek, W. Fukarek: Thin Solid Films **351**, 151 (1999)
289. H. Biederman, V. Studzia, D. Slavinska, J. Glosik: Vacuum **52**, 415 (1999)
290. J.T. Harnack, F. Thieme, C. Benndorf: Surf. Coat. Technol. **39/40**, 285 (1989)
291. S.J. Harris, A.M. Weiner, W.J. Meng: Wear **211**, 208 (1997)
292. S.J. Harris, A.M. Weiner, C.H. Olk, M. Grischke: Wear **219**, 219 (1998); S.J. Harris, A.M. Weiner, M. Grischke: Surf. Coat. Technol. **120**, 561 (1999);
293. C. Benndorf, M. Grischke, H. Koeberle, R. Memming, A. Brauer, F. Thieme: Surf. Coat. Technol. **39/40**, 275 (1989)
294. H. Koeberle, M. Grischke, F. Thieme, C. Benndorf, R. Memming: Surf. Coat. Technol. **39/40**, 275 (1989)
295. K. Bewilogua, C.V. Cooper, C. Specht, J. Schroeder, R. Wittorf, M. Grischke: Surf. Coat. Technol. **127**, 224 (2000)
296. M. Grischke, K. Bewilogua, K. Trojan, H. Dimigen: Surf. Coat. Technol. **74–75**, 739 (1995)
297. D. Dalacu, L. Martinu: J. Appl. Phys. **87**, 228 (2000)
298. W. Dauer, G. Betz, H. Bangert, A. Borguur, C. Eisenmenger-Sittner: J. Vac. Sci. Technol. A **12**, 3157 (1994)
299. H. Biederman: Vacuum **37**, 367 (1987)
300. L. Martinu, H. Biederman: J. Zemek, Vacuum **35**, 171 (1985)
301. H. Biederman, I. Chudacek, D. Slavinska, L. Martinu, J. David, S. Nespurek: Vacuum **39**, 13 (1989)
302. H. Biederman, Z. Chmel, A. Fejvar, M. Misina, J. Pesicka: Vacuum **40**, 377 (1990)
303. K. Kashiwagi, Y. Yoshida, Y. Murayama: J. Vac. Sci. Technol. A **5**, 1828 (1987)
304. J. Werner, A. Heilmann, V. Hopfe, F. Homilius, B. Steiger, O. Stenzel: Thin Solid Films **237**, 193 (1994)
305. A. Heilmann, G. Kampfrath, V. Hopfe: J. Phys. D: Appl. Phys. **21**, 986 (1988)
306. A. Heilmann, C. Hamann, Do Ngoc Uan: Phys. Stat. Sol. (a) **110**, K29 (1988); A. Heilmann, C. Hamann, G. Kampfrath, Do Ngoc Uan: Phys. Stat. Sol. (a) **114**, 551 (1989)

307. A. Heilmann, C. Hamann, G. Kampfrath, V. Hopfe: Vacuum **41**, 1472 (1990)
308. A. Heilmann, J. Werner, V. Hopfe: Z. Phys. D **26**, S39 (1993)
309. A. Heilmann, C. Hamann: Prog. Coll. Polym. Sci. **85**, 102 (1991)
310. A. Heilmann, J. Werner, O. Stenzel, F. Homilius: Thin Solid Films **246**, 77 (1994)
311. A. Heilmann, C. Hamann, G. Kampfrath: in *Adv. in Low-Temperature Plasma Chemistry, Technology, Applications* **Bd. 3** H.V. Boenig (Ed.), Technomic Pub. Lancaster (1991) S. 202
312. A. Heilmann, C. Hamann: Springer Series in Solid-State Sciences **107**, 429 (1992)
313. A. Heilmann, A.-D. Müller, J. Werner: Surf. Rev. Lett. **3**, 1113 1996
314. A. Heilmann, J. Werner, D. Schwarzenberg, S. Henkel, P. Grosse, W. Theiß: Thin Solid Films **270**, 103 (1995)
315. A. Heilmann, A.-D. Müller, J. Werner, F. Müller: Thin Solid Films **270**, 371 (1995)
316. A.V. Savchuk, E.N. Salkova, T.A. Sergan, M.S. Soskin, A. Heilmann: SPIE Proc. **2108**, 65 (1993); A.V. Savchuk, E.N. Salkova, T.A. Sergan, M.S. Soskin, A. Heilmann: Opt. Memory Neural Networks **2**, 205 (1993)
317. F. Müller, A.-D. Müller, O. Meissner, A. Heilmann, M. Hietschold: Rev. Sci. Inst. **68**, 3104 (1997)
318. A. Heilmann, J. Werner: Thin Solid Films **317**, 21 (1998)
319. Do Ngoc Uan, A. Heilmann: Vacuum **49**, 51 (1998)
320. A. Heilmann, M. Quinten, J. Werner: Europ. J. Phys. B **3**, 455 (1998)
321. A. Heilmann, M. Quinten, A. Kiesow: MRS Proc. **501**, 73 (1998)
322. M. Quinten, A. Heilmann, A. Kiesow: Appl. Phys. B **68**, 707 (1999)
323. A. Heilmann, U. Kreibig: European Phys. J. Appl. Phys. **10**, 193 (2000)
324. A. Heilmann, A. Kiesow, M. Gruner, U. Kreibig: Thin Solid Films **343–344**, 175 (1999)
325. A.V. Savchuk, E.N. Salkova, T.A. Sergan, A. Heilmann: Phys. Stat. Sol. (a) **122**, K83 (1990)
326. A.-D. Müller, A. Heilmann, F. Müller: Thin Solid Films, **281–282**, 112 (1996)
327. Y. Asano: Thin Solid Films **105**, 1 (1983)
328. G. Kampfrath, A. Heilmann, C. Hamann: Vacuum **38**, 1 (1988)
329. J. Werner, A. Heilmann, F. Müller: Appl. Phys. Lett. **66**, 3426 (1995)
330. A. Heilmann, J. Werner, F. Homilius, F. Müller: J. Adhesion Sci. Technol. **9**, 1181 (1995)
331. A. Heilmann, F. Homilius, J. Werner: in *Industrial Applications of Plasma Physics* (ISPP 13), G. Bonizzoni, W. Hooke, E. Sindoni (Eds.), Editrice Compositori, Bologna, S. 563 (1992)
332. F. Homilius, A. Heilmann, J. Werner, W. Grünewald: Phys. Stat. Sol. (a), **137**, 145 (1993)
333. J. Werner, A. Heilmann, O. Stenzel: Int. J. Electron. **77**, 945 (1994)
334. W. Grünewald, A. Heilmann, F. Homilius: Fresenius J. Anal. Chem. **349**, 238 (1994); W. Grünewald, A. Heilmann, C. Reinhardt: Appl. Surf. Sci. **93**, 157 1996
335. M. Hietschold, W. Vollmann, A. Mrwa, A. Heilmann, P.K. Hansma: Phys. Stat. Sol. (a) **131**, 59 (1992); M. Hietschold, O. Pester, W. Vollmann, A. Heilmann, P. Staebeler, H. Sbosny X. Grahlert, H.-U. Sonntag, A. Bruska, B. Winzer, T. Schimmel, L. Koenders: *Atomic Force Microscopy/Scanning Tunneling Microscopy*, S.H. Cohen (Ed.), Plenum Press, New York (1994) S. 143
336. C. Reinhardt, A. Heilmann, W. Grünewald, C. Hamann: Thin Solid Films **235**, 57 (1993)

337. D. Salz, B. Mahltig, A. Baalmann, M. Wark, N. Jaeger: Phys. Chem. Chem. Phys. **2**, 3105 (2000)
338. R. Lamber, S. Wetjen, G. Schulz-Ekloff, A. Baalmann: J. Phys. Chem. **99**, 13834 (1995)
339. R. Lamber, A. Baalmann, N.I. Jaeger, G. Schulz-Ekloff, S. Wetjen: Adv. Mater. **6**, 224 (1994)
340. D. Salz, R. Lamber, M. Wark, A. Baalmann, N. Jaeger: Phys. Chem. Chem. Phys. **1**, 4447 (1999)
341. P. Stulik, H. Biederman, D. Slavinska, A. Fejfar, I. Chudacek, P. Mackus: Int. J. Electron. **70**, 509 (1991)
342. A. Heilmann, A. Kiesow, E. Hoinkis, BENSC Experimental Reports 1997, p. 253
343. G.J. Kovacs, P.S. Vincett, C. Tremblay, A.L. Pundsack: Thin Solid Films **101**, 23 (1983)
344. X. Yu, P.M. Duxbury, G. Jeffers, M.A. Dubson: Phys. Rev. B **44**, 13163 (1991)
345. T.L. Morkved, P. Wiltzius, H.M. Jaeger, D.G. Grier, T.A. Witten: Appl. Phys. Lett. **64**, 422 (1994)
346. C. Bubeck: Thin Solid Films **178**, 483 (1989)
347. S. Norrman, T. Andersson, C.G. Granqvist, O. Hunderi: Phys. Rev. B **18**, 674 (1978)
348. C.G. Granqvist, O. Hunderi: Phys. Rev. B **16**, 3513 (1977)
349. P.G. Borziak, Y.A. Kulyupin, S.A. Nepijko, V.G. Shamonya: Thin Solid Films **76**, 359 (1981)
350. L.L. Kazmerski, D.M. Racine: J. Appl. Phys. **46**, 791 (1975)
351. C.G. Granqvist, R.A. Buhrman: J. Appl. Phys. **47**, 2200 (1976)
352. S. Regnier, M. Gillet: Z. Phys. D **19**, 311 (1991)
353. K. Kimura: Bull. Chem. Soc. Jpn. **60**, 3093 (1987)
354. A. Sukawara, Y. Nakamura, O. Nittono: J. Cryst. Growth **99**, 583 (1990)
355. V. Haas, R. Birringer: Nanostruct. Mat. **1**, 491 (1992)
356. Eds. J.F. Moulder, W.F. Stickle, P.E Sobol, K.D. Bombden, Perkin Elmer, Eden Prairie: Handbook of X-Ray Photoelectron Spectroscopy (1992)
357. A. Heilmann, M. Kelly, E. Kay: unpublished results
358. S. Giorgio, J. Urban, W. Kunath: Phil. Mag. A **60**, 553 (1989)
359. T. Kizuka, T. Kachi, N. Tanaka: Z. Phys. D **26**, S58 (1993)
360. S. Iijima, T. Ichihashi: Phys. Rev. Lett. **56**, 616 (1986)
361. S.W. Link, C. Burda, Z.L. Wang, M.A. El-Sayed: J. Chem. Phys. **111**, 1255 (1999)
362. H. Hofmeister, W.G. Drost, A. Berger: Nanostruct. Mater. **12**, 207 (1999)
363. D.J. Smith, A.K. Petford-Long, L.R. Wallenberg, J.O. Boivin: Science **233**, 872 (1986)
364. A.N. Patil, D.Y. Paintankar, N. Otsuka, R.P. Andres: Z. Phys. D **26**, 135 (1993)
365. D. Narayanaswamy, L.D. Marks: Z. Phys. D **26**, S70 (1993)
366. R. Uyeda: in *Morphology of Crystals*, I. Sunagawa (Ed.), Kluwer, Dordrecht (1987) S. 367
367. J.O. Malm, G. Schmidt, B. Morun: Phil Mag. A **63**, 487 (1991)
368. ASTM-Datei Nr. 50642 (JCPDS-ICDD)
369. J.O. Bovin, J.O. Malm: Z. Phys. D **19**, 293 (1991)
370. G. Reiners: Thin Solid Films **143**, 311 (1986)
371. R.J. Warmack, S.L. Humphrey: Phys. Rev. B **34**, 2246 (1986)
372. M.S. Kunz, K.R. Shull, A.J. Kellock: J. Appl. Phys. **72**, 4458 (1992)
373. K. Yata, T. Yamaguchi: J. Am. Cer. Soc. **75**, 2071 (1992); **75**, 2910 (1992)
374. F. Faupel: phys. stat. sol. (a) **134**, 9 (1992)

375. F. Faupel, R. Willecke, A. Thran: Mat. Sci. Engn. **R22**, 1 (1998)
376. W. Ostwald: Z. Physik. Chem. **34**, 495 (1900)
377. I.M. Lifshitz, V.V. Slyozov: J. Phys. Chem. Solids **19**, 35 (1961)
378. C. Wagner: Z. Elektrochemie **65**, 581 (1961)
379. J.W. Martin, R.D. Doherty: *Stability of Microstructures in Metallic Systems*, Cambridge University Press, Cambridge (1976)
380. A.J. Ardell: Acta Metall. **20**, 61 (1972)
381. P.W. Voorhees, M.E. Glicksman: Acta Metall. **32**, 2001 (1984); **32**, 2013 (1984)
382. A.D. Brailsford, P. Wynblatt: Acta. Metall. **27**, 489 (1979)
383. C.K.L. Davies, P. Nash, R.N. Stevens: Acta Metall. **28**, 179 (1980)
384. Y. Enomoto, M. Tokuyama, K. Kawasaki: Acta Metall. **34**, 2119 (1986); Y. Enomoto, K. Kawasaki: Acta Metall. **37**, 1399 (1989)
385. T. Kuepper, N. Masbaum: Acta Metall. Mater. **42**, 1847 (1994)
386. S.C. Yang, G.T. Higgins, P. Nash: Mat. Sci. Technol. **8**, 10 (1992)
387. W. Bender, L. Ratke: Z. Metallkunde **83**, 541 (1992)
388. S.S. Kang, D.N. Yoon: Metall. Trans. A **13**, 1405 (1982)
389. D. Uffelmann, W. Bender, L. Ratke, B. Ferbacher: Acta Metall. Mater. **43**, 173 (1995)
390. L. Radke: Progr. Astronautics Aeoronautics **130**, 661 (1990)
391. S. Mantl: Mater. Sci. Rep. **8**, 1 (1995)
392. P.F.P. Fichtner, W. Jäger, K. Radermacher, S. Mantl: Nucl. Instr. Meth. Phys. Res. B **59/60**, 632 (1991); H. Trinkaus, S. Mantl: Nucl. Instr. Meth. Phys. Res. B **80/81**, 862 (1991); P.F.P. Fichtner, H. Schroeder, H. Trinkaus: Acta Metall. Mater. **39**, 1845 (1991)
393. F.M. Mirabella, J.S. Barley: J. Polym. Sci. B **32**, 2187 (1994)
394. J. Rothaut, H. Schröder, H. Ullmaier: Phil. Mag. A **47**, 781 (1983)
395. K. Yata, T. Yamaguchi: J. Mater. Sci. **25**, 101 (1992)
396. T.H. Courtney: Metall. Trans. A **8**, 671 (1977); **8**, 679 (1977); **8**, 685 (1977)
397. S. Maeda, S. Iwabuchi, M. Shiojiri: Jap. J. Appl. Phys. **23**, 830 (1984)
398. D.W. Pashley, M.J. Stowell, M.H. Jacobs, T.J. Law: Phil. Mag. **10**, 127 (1964), D.W: Pashley: Adv. Phys. **14**, 327 (1965)
399. V.B. Jipson, H.N. Lynt, J.F. Graczyk: Appl. Phys. Lett. **43**, 27 (1983)
400. M. Flüeli, P.A. Buffat, J.P. Borel: Surf. Sci. **202**, 343 (1988)
401. B.C. Gates, L. Guczi, H. Knozingen: *Metal Clusters in Catalysis*, Elsevier, New York (1986)
402. A.T. Frommhold: *Theory of Metal Oxidation*, North-Holland, Amsterdam (1980)
403. R. Gatt, G.A. Niklasson, M. Heim, C.G. Granqvist: SPIE Proc. **2017**, 338 (1993)
404. R. Okada, S. Iijima: Appl. Phys. Lett. **58**, 1662 (1991)
405. S. Sako, K. Ohshima, T. Fujita: J. Phys. Soc. Japan **59**, 662 (1990)
406. S. Hayashi, S. Kawata, H.M. Kim, K. Yamamoto: Jap. J. Appl. Phys. A **32**, 4870 (1993)
407. M. Rauh, P. Wišmann: Thin Solid Films **228**, 121 (1993)
408. M. Takagi: J. Phys. Soc. (Japan) **9**, 359 (1954)
409. M.Y. Zhou, P. Sheng: Phys. Rev. B **43**, 3460 (1991)
410. R. Kofman, P. Cheyssac, R. Garrigos, Y. Lereah, G. Deutscher: Z. Phys. D **20**, 267 (1991)
411. K.M. Unruh, B.M. Patterson, S.I. Shah: in *Physical Phenomena in Granular Materials*, C.D. Cody, T.H. Geballe, P. Sheng (Eds.), MRS Symp. Proc. **195** (1990) S. 567
412. S. Valkealahti, M. Manninen: Z. Phys. D **26**, 255 (1993)

413. T. Castro, R. Reifernberger, E. Choi, R. P. Andres: Phys. Rev. B **42**, 8548 (1990)
414. D.R. Lide, H.P:R. Frederikse (Eds.): *CRC Handbook of Chemistry and Physics*, CRC Boca Raton (1995)
415. T. Ohashi, K. Kuroda, H. Saka: Phil. Mag. B **65**, 1041 (1992)
416. D.K. Hale: J. Mater. Sci. **11**, 2105 (1976)
417. M. Avalos-Borja: PhD Thesis, Stanford University (1983)
418. V.P. Safonov, V.M. Shalaev, V.A. markel, Y.E. Danilova, N.N. Lepeshkin, W. Kim, S.G. Rautian, R.L. Arnstrong: Phys. Rev. Lett. **80**, 1102 (1998)
419. M. Kaempfe, T. Rainer, K.J. Berg, G. Seifert, H. Graener: Appl. Phys. Lett. **74**, 1200 (1999)
420. G. Seifert, M. Kaempfe, K.J. Berg, H. Graener: Appl. Phys. B **71**, 795 (2000); Appl. Phys. B **71**, 355 (2001);
421. H.G. Craighead, R.E. Howard, P.F. Liao, D.M. Tennant, J.E. Sweeney: Appl. Phys. Lett. **40**, 663 (1982)
422. S.Y. Suh, H.G. Craighead: Appl. Opt. **24**, 208 (1985)
423. H.G. Craighead, R.E. Howard, J.E. Sweeney, D.M. Tennant: J. Vac. Sci. Technol. **20**, 316 (1982)
424. B.G. Atabaev, E.M. Dubinina, S.S. Elovikov, M.M. Saad Eldin: Thin Solid Films **141**, 137 (1986)
425. H. Zhu, R.S. Averback: Phil Mag. Lett. **73**, 27 (1996)
426. M. Miki-Yoshida, S. tehuacanero, M. Jose-Yacaman: Surf. Sci. Lett. **274**, L569 (1992)
427. M. Eßner, K. Morgenstern, G. Rosenfeld, G. Comsa: Surf. Sci. **402-404**, 341 (1998)
428. M. Yeadon, J.C. Yang, R.S. Averback, J.W. Bullard, J.M. Gibosn: Nanostruct. Mat. **10**, 731 (1998)
429. A. Heilmann, J. Erben, M. Plischke: unpublished
430. A. Heilmann, J. Werner, D. Schwarzenberg: unpublished
431. L.I. Maissel, R. Glang: *Handbook of Thin Film Technology*, McGraw-Hill, New York (1995)
432. D.S. McLachlan: J. Phys. C: Solid State Phys. **18**, 1891 (1985); **19**, 1339 (1986)
433. A. Heilmann, PhD Thesis, TU Karl-Marx-Stadt (1988)
434. K. Fuchs: Proc. Cambridge Phil. Soc. **34**, 100 (1938)
435. E.H. Sondheimer: Adv. Phys. **1**, 1 (1952)
436. A.P. Dorey, J. Knight: Thin Solid Films **4**, 445 (1969)
437. U. Jacob, J. Vancea, H. Hoffmann: Phys. Rev. B **41**, 11852 (1990); G. Reiss, E. Hastreiter, H. Brückl, J. Vancea: Phys. Rev. B **43**, 5176 (1991)
438. F. Abeles, M.L. Theye: Phys. Lett. **4**, 348 (1963)
439. U. Jacob, J. Vancea, H. Hoffmann: J. Phys.: Cond. Matt. **1**, 9867 (1989)
440. D. Stauffer, A. Aharony: Einführung in die Perkolationstheorie, VCH Weinheim (1995)
441. D. Stauffer: Phys. Rep. **54**, 1 (1979)
442. S. Kirkpatrick: Phys. Rev. Lett. **27**, 1722 (1971); Rev. Mod. Phys. **54**, 574 (1974)
443. I. Webman, J. Jortner, M.H. Cohen: Phys. Rev. B **11**, 2885 (1975)
444. D.J. Bergman: Physica A **157**, 72 (1989)
445. S.P. McAlister, A.D. Inglis, P.M. Kayll: Phys. Rev. B **31**, 5113 (1985)
446. E.B. Priesley, B. Abeles, R.W. Cohen: Phys. Rev. B **12**, 2121 (1975)
447. D.M. Grannan, J.C. Garland, D.B. Tanner: Phys. Rev. B **46**, 375 (1981)
448. I.G. Chen, W.B. Johnson: J. Mater. Sci. **21**, 3162 (1986)
449. R.F. Voss, R.B. Laibowitz, E.I. Allessandrini: Phys. Rev. Lett. **49**, 1441 (1982)

450. M. Octavio, G. Gutierrez, J. Aponte: Phys. Rev. B **3**, 2461 (1987)
451. A. Palevski, G. Deutscher: Phys. Rev. B **34**, 431 (1986)
452. L. Cheriet, H.H. Helbig, S. Araja: Phys. Rev. B **39**, 9828 (1989)
453. R.M. Hill: Proc. Roy. Soc. A **309**, 377 (1969); A **309**, 797 (1969)
454. C.A. Neugebauer, M.B. Webb: J. Appl. Phys. **33**, 74 (1962)
455. J.G. Swanson, D.S. Campbell, J.C. Anderson: Thin Solid Films **1**, 325 (1967)
456. T. Andersson: J. Appl. Phys. **47**, 1752 (1976)
457. M. Gazicki, H. Yasuda:
458. K.D. Leaver: J. Phys. C: Solid State Phys. **10**, 249 (1977)
459. C.J. Adkins: J. Phys. C: Solid State Phys. **20**, 235 (1987) J. Appl. Polym. Sci.: Appl. Polym. Symp. **38**, 35 (1984)
460. G. Dittmer: Thin Solid Films **9**, 317 (1972)
461. M. Mostefa: Solid State Comm. **73**, 365 (1990)
462. B. Abeles: in *Applied Solid State Science* 6, R. Wolfe (Ed.), Academic Press, New York (1976) S. 1
463. J.E. Morris, A. Mello, C.J. Adkins: in *Physical Phenomena in Granular Materials*, C.D. Cody, T.H. Geballe, P. Sheng (Ed.), MRS Symp. Proc. **195**, 181 (1990)
464. B. Barwinski: Thin Solid Films **128**, 1 (1985)
465. C.J. Adkins in: Physical Phenomena in Granular Materials, Ed. C.D. Cody, T.H. Geballe, P. Sheng, MRS Symp. Proc. **195**, S. 223 (1990)
466. P. Sheng, B. Abeles, Y. Arie: Phys. Rev. Lett. **31**, 44 (1973)
467. J.E. Morris, T.J. Coutts: Thin Solid Films **47**, 3 (1977)
468. C.J. Adkins: J. Phys.: Condens. Matter **1**, 1253 (1989)
469. S.H. Kwan, F.G. Shin, L. Tsui: J. Mater. Sci. **15**, 2978 (1980)
470. R. Blessing, H. Pagnia: Thin Solid Films **52**, 333 (1975); H. Pagnia, N. Sotnik: Phys. Stat. Sol. (a) **108**, 11 (1988)
471. M.I. Newton, G. McHale, P.D. Hooper, M.R. Willis, S.D. Burt: Vacuum **46**, 315 (1995)
472. A. Panckow, T.P. Drüsedau, U. Schmidt: MRS Conf. Proc. **903**, 461 (1994); A. Panckow, T.P. Drüsedau, B. Schroeder: MRS Conf. Proc. **897**, 571 (1995)
473. G.G. Paulson, A.L. Friedberg: Thin Solid Films **5**, 47 (1970)
474. K. Narayandas, M. Radhahrishnan, C. Balasubramanian: Thin Solid Films **67**, 357 (1980)
475. D. Schlemminger, D. Stark: Thin Solid Films **137**, 49 (1986); D. Schumacher, D. Stark: Thin Solid Films **139**, 33 (1986)
476. D. Hecht, D. Stark: Thin Solid Films **238**, 258 (1994)
477. P. Bieganski, E. Dorbierzewska-Mozrymas Int. J. Electronics **73**, 965 (1992)
478. A. Chatterjee, D. Chakravorty: J. Phys. D.: Appl. Phys. **23**, 1097 (1990)
479. H. Raether: *Surface Plasmons*, Springer Tracts Mod. Phys. **111**, Springer, Berlin (1988)
480. P. Kloucek (Ed.) *Handbook of Infrared materials* Marcel Dekker, New York (1991)
481. J. Babiskin, J.R. Anderson (Eds.): *American Institute of Physics Handbook*, McGraw-Hill, New York (1972)
482. C.G. Granqvist, O. Hunderi: Phys. Rev. B **18**, 2897 (1978); C.G. Granqvist: Z. Physik B **30**, 29 (1978)
483. F. Abeles, Y. Borensztein, T. Lopez-Rios: Festkörperprobleme **14**, 93 (1984)
484. G. Mie: Ann. Phys. **25**, 377 (1908)
485. M. Quinten: PhD Thesis, Universität des Saarlandes Saarbrücken (1989)
486. S. Asano, G. Yamamoto: Appl. Opt. **14**, 29 (1975)
487. W. Seitz: Ann. Phys. **21**, 1013 (1906)
488. W. von Ignatowski: Ann. Phys. **18**, 495 (1905)

489. U. Kreibig: J. Phys. F: Metal. Phys. **4**, 999 (1974)
490. II. Ehrenreich, D.R. Philipp: Phys. Rev. **128**, 1622 (1962)
491. B.K. Russell, J.G. Mantovani, V.E. Anderson, R.J. Warmack, T.L. Ferrell: Phys. Rev. B **35**, 2151 (1987)
492. S. Yamaguchi: J. Phys. Soc. Japan **15**, 1577 (1960); J. Phys. Soc. Japan **18**, 266 (1962)
493. R.W. Tokarsky, J.P. Marton: J. Appl. Phys. **45**, 3051 (1974)
494. S.L. McCarthy: J. Vac. Sci. Technol. **13**, 135 (1976)
495. Y. Borensztein, P. DeAndres, R. Monreal, T. Lopez-Rios, F. Flores: Phys. Rev. B **33**, 2828 (1986)
496. M.H. Lee, P.J. Dobson, B. Cantor: Thin Solid Films **219**, 199 (1992)
497. E. Anno, R. Hoshino: J. Phys. Soc. Japan **50**, 1209 (1981); E. Anno, R. Hoshino: Surf. Sci. **144**, 567 (1984); E. Anno: J. Opt. Soc. Am. B **3**, 194 (1986)
498. W. Gotschy, K. Vonmetz, A. Leitner, F.R Aussenegg: Opt. Lett. **21**, 1099 (1996); Appl. Phys. B **63**, 381 (1996)
499. T. Wenzel, J. Bosbach, F. Stietz, F. Träger: Surf. Sci. **432**, 257 (1999)
500. E. Anno, R. Hoshino: J. Phys. Soc. Japan **51**, 1185 (1982)
501. K.D. Cummings, J.C. Garland, D.B. Tanner: Phys. Rev. B **30**, 4170 (1984)
502. N. Emeric, A. Emeric: Thin Solid Films **1**, 13 (1967)
503. T. Yamaguchi, S. Yoshida, A. Kinbara: Thin Solid Films **21**, 173 (1974)
504. J.P. Goudonnet, J.L. Bijeon, M. Pauty: Thin Solid Films **177**, 49 (1989)
505. U. Kreibig, A. Althoff, H. Pressmann: Surf. Sci. **106**, 308 (1981)
506. U. Kreibig, B. Schmitz, H.D. Breuer: Phys. Rev. B **36**, 5027 (1987)
507. U. Kreibig, L. Genzel: Surf. Sci. **156**, 678 (1985)
508. U. Kreibig: Z. Phys. B **31**, 39 (1978)
509. U. Kreibig: Z. Phys. D **3**, 239 (1986)
510. F.R. Aussenegg, A. Leitner, H. Gold: Appl. Phys. A **60**, 97 (1995)
511. F. Hache, D. Ricard, C. Flytzanis, U. Kreibig: Appl. Phys. A **47**, 347 (1988)
512. F. Hache, D. Ricard, C. Flytzanis: J. Opt. Soc. Am. B 3, 1647 (1986)
513. K. Kadono, T. Sakaguchi, M. Miya, J. Matsuoka, T. Fukumi, H. Tanaka: J. Mat. Sci.: Mat. Electron. 4, 59 (1993)
514. K. Fukumi, A. Chayahara, K. Kadono, T. Sakaguchi, Y. Horino, M. Miya, K. Fujii, J. Hayakawa, M. Satou: J. Appl. Phys. **75**, 3075 (1994)
515. T. LaPeruta, E.A. van Wagenen, J.J. Roche, P. Kitipichai, G.E. Wnek, G.M. Korenowski: SPIE Proc. **1497**, 357 (1991)
516. M.G. Kuzyk, M.P. Andrews, U.C. Paek, C.W. Dirk: SPIE Proc. **1560**, 44 (1991)
517. M. van Exter, A. Lagendijk: Phys. Rev. Lett. **60**, 49 (1988)
518. D. Steinmüller-Nethl, R. A. Höpfel, F. Gornik, A. Leitner, F.R. Aussenegg: Phys. Rev. Lett. **68**, 389 (1992);
519. B. Lamprecht, A. Leitner, F.R. Aussenegg: Appl. Phys. B **64**, 269 (1997)
520. H. Inouye, K. Tanaka, I. Tanahashi, K. Hirao: Phys. Rev. B **57**, 11334 (1998)
521. V. Halte, J. Guille, J.-C. Merle, I. Perakis, J.-Y. Bigot: Phys. Rev. B **60**, 11738 (1999); V. Halte, J.-Y. Bigot, B. Palpant, M. Broyer, B. Prevel, A. Perez: Appl. Phys. Lett. **75**, 3799 (1999); J.-Y. Bigot, V. Halte, J.-C. Merle, A. Daunois: Chem. Phys. **252**, 181 (2000)
522. N. Del Fatti, C. Flytzanis, F. Vallée, Appl. Phys. B **68**, 433 (1999); N. Del Fatti, F. Vallée, C. Flytzanis, Y. Hamanaka, A. Nakamura: Chem. Phys. **251**, 215 (2000); C. Voisin, D. Christofilos, N. Del Falli, F. Vallée, B. Prevel, E Cottanicin, J. Lerme, M. Pellarin, M. Broyer: Phys. Rev. Lett. **85**, 2200 (2000)

523. M. Perner, S. Gresillon, J. März, G. v. Plessen, J. Feldmann, J. Porstendorfer, K.-J. Berg, G. Berg: Phys. Rev. Lett. **85**, 792 (2000); M. Perner, T. Klar, S. Grosse, U. Lemmer, G. v. Plessen, W. Spirkl, J. Feldmann: J. Luminesc. **76–77**, 181 (1998); T. Klar, M. Perner, S. Grosse, G. v. Plessen, W. Spirkl, J. Feldmann: Phys. Rev. Lett. **80**, 4249 (1998)
524. J. Lehmann, M. Merschdorf, W. Pfeiffer, A. Thon, S. Voll, G. Gerber: Phys. Rev. Lett. **85**, 2921 (2000)
525. S.I. Lee, T.W. Noh, J.R. Gaines, Y.H. Ko; E.R. Kreidler: Phys. Rev. B **37**, 2918 (1988)
526. K. Baba, K. Hayashi, I. Syuaib, K. Yamaki, M. Miyagi: Appl. Opt. **36**, 2421 (1997); K. Baba, Y. Ohkuma, T. Yonezawa, M. Miyagi: Appl. Opt. **40**, 2796 (2001)
527. J. Bosbach, D. Martin, F. Stietz, T. Wenzel, F. Träger: Appl. Phys. Lett. **74**, 2605 (1999)
528. F. Stietz, F. Träger: Phil. Mag. B **79**, 1281 (1999)
529. W.D. Bragg, K. Banerjee, V.A. Podolskiy, V.P. Safonov, J.G. Zhu, V.M. Shalaev, Z.C. Ying: SPIE Proc. **3791**, 85 (1999)
530. J. Martin, A. Kiesow, A. Heilmann, R. Wannemacher: Appl. Opt. **40**, 5726 (2001)
531. J. Werner, A. Heilmann: unpublished
532. H. Hovel, S. Fritz, A. Hilger, U. Kreibig, M. Vollmer, Phys. Rev. B. **48**, 18178 (1993)
533. S. Jain, N. Arora: J. Phys. Chem. Sol. 35, 1231 (1974)
534. S.D. Stookey, G.H. Beall, J.E. Pierson: J. Appl. Phys. **49**, 5114 (1978)
535. J. Porstendorfer, K.J. Berg, G. Berg: Journal of Quant. Spectr. and Radiative Transfer **63**, 479 (1999)
536. R.E. Hummel: Optische Eigenschaften von Metalllen und Legierungen, Springer Berlin (1971)
537. T. Kawaguchi, S. Maruno, K. Masui: J. Non-Crystall. Sol. **95/96**, 777 (1987); T. Kawaguchi, K. Masui: Jap. J. Appl. Phys. **26**, 15 (1987)
538. Y. Yagil, G. Deutscher: Thin Solid Films **152**, 465 (1987)
539. U. Kreibig, C. von Fragstein: Z. Phys. **224**, 307 (1969)
540. E.D. Palik (Ed.): *Handbook of Optical Constants of Solids*, Academic Press, Orlando (1985)
541. P.B. Johnson, R.W. Christy: Phys. Rev. B 6, 4370 (1972); Phys. Rev. B **9**, 5056 (1974)
542. H.J. Hagemann, W. Gudat, C. Kunz: DESY-Report SR-74/7 Hamburg (1974)
543. R. Gans: Ann. Phys. **37**, 881 (1912)
544. R. Fuchs: Phys. Rev. B **11**, 1732 (1975)
545. M. Quinten, U. Kreibig: Appl. Opt. **32**, 6173 (1993)
546. M. Quinten: Z. Phys. B **101**, 211 (1996)
547. G. Banhegyi: Colloid Polymer Sci. **264**, 1030 (1986)
548. D.K. Das-Gupta, K. Doughty: Thin Solid Films **158**, 93 (1988)
549. F. Brouers: J. Phys. C **19**, 7183 (1986)
550. G. Ondracek: Z. Metallk. **77**, 603 (1986); Mater. Chem. Phys. **15**, 281 (1986)
551. W. Kreher, W. Pompe: Internal Stress in Heterogeneous Solids, Akademie-Verlag Berlin (1989)
552. D.E. Aspnes: Thin Solid Films **89**, 249 (1982)
553. J.C. Maxwell Garnett: Phil. Trans. Royal Society London **203**, 385 (1904); **205**, 237 (1906)
554. J.W. Lord Rayleigh: Phil. Mag. **34**, 481 (1892)
555. R. Ruppin: Phys. Stat. Sol. (b) **87**, 619 (1978)
556. D. Polder, J.H. van Santen: Physica XII, 257 (1946)

557. H. Fricke: Phys. Rev. **24**, 575 (1924)
558. A.H. Shivola, I.V. Lindell: Electr. Lett. **26**, 119 (1990)
559. G.B. Smith: Optics Comm. **71**, 279 (1989)
560. D.A.G. Bruggeman: Annalen der Physik **24**, 636 (1935)
561. Z. Hashin, S. Shtrikman: J. Appl. Phys. **33**, 3125 (1962)
562. D.J. Bergman: Phys. Rev. B **23**, 3058 (1981); Phys. Rep. (Phys. Lett. C) **43**, 377 (1978)
563. G.W. Milton: Appl. Phys. Lett. **37**, 300 (1980); J. Appl. Phys. **52**, 5286 (1980)
564. O. Wiener: Abh. der sächs. Ges. der Wissenschaften, math.-phys. Kl. **32**, 509 (1912)
565. F. Lado, S. Torquato: Phys. Rev. B **33**, 3370 (1986); S. Torquato, F. Lado: Phys. Rev. B **33**, 6428 (1986); S. Torquato: J. Chem. Phys. **84**, 6345 (1986)
566. Y. Kantor, D.J. Bergman: J. Phys. C: Sol. Stat. Phys. **15**, 2033 (1982)
567. H. Looyenga: Physica **31**, 401 (1965)
568. K. Lichtenecker: Physik. Z. **27**, 117 (1926)
569. J. Monecke: J. Phys. C: Cond. Matt. **6**, 907 (1994)
570. P. Sheng: Phys. Rev. Lett. **45**, 60 (1980); Phys. Rev. B **22**, 6364 (1980)
571. B.N.J. Persson, A. Liebsch: Sol. State Comm. **44**, 1637 (1982); A. Liebsch, B.N.J. Persson: J. Phys. C. Sol. State Phys. **17**, 5375 (1983); A. Liebsch, P.V. Gonzalez: Phys. Rev. B **29**, 6907 (1983)
572. W. Lamb, D.M. Wood, N.W. Ashcroft: Phys. Rev. B **21**, 2248 (1980)
573. V.A. Davis, L. Schwartz: Phys. Rev. B **31**, 5155 (1985); Phys. Rev. B **33**, 6627 (1986)
574. P. Chylek, V. Srivastava: Phys. Rev. B **27**, 5098 (1983)
575. D. Bedeaux, J. Vlieger: Thin Solid Films **102**, 265 (1983)
576. S. Fend, M.F. Thorpe, E.J. Garboczi: Phys. Rev. B **31**, 276 (1985); E.J. Garboczi, M.F. Thorpe: Phys. Rev. B **31**, 7276 (1985); E.J. Garboczi, M.F. Thorpe: Phys. Rev. B **33**, 3289 (1986)
577. T. Robin, B. Souillard: Physica A **193**, 779 (1993)
578. X.C. Zeng, D.J. Bergman, P.M. Hui, D. Stoud: Phys. Rev. B **38**, 10970 (1988)
579. M. Gomez, L. Fonseca: Thin Solid Films **125**, 243 (1985); M. Gomez, L. Fonseca, G. Rodriguez, A. Velazquez, L. Cruz: Phys. Rev. B **32**, 3429 (1985)
580. R.G. Barrera, G. Monsivais, W.L. Mochan, E. Anda: Phys. Rev. B **39**, 9998 (1989)
581. A.V. Cherkaev, L.V. Gibiansky: Proc. Roy. Soc. Edinburgh **122A**, 93 (1992)
582. G.A. Niklasson, C.G. Granqvist, O. Hunderi: Appl. Opt. **20**, 26 (1981); G.A. Niklasson, C.G. Granqvist: Sol. Energy Mat. **5**, 173 (1981); G.A. Niklasson: Sol. Energy Mat. **17**, 217 (1988)
583. G.I. Hornyak, K.L.N. Phani, D.L. Kunkel, V.P. Menon, C.R. Martin: Nanostruct. Mat. **6**, 839 (1995)
584. G.L. Hornyak, C.J. Patrissi, C.R. Martin: J. Phys. Chem. B **101**, 1548 (1997)
585. V.N. Rai: Pramana J. Phys. **31**, 313 (1988)
586. K. Watanabe, K. Kobayashi, C.C. Wong, Y.M. Xiong, T. Saitoh, F. Hyuga: Thin Solid Films **270**, 97 (1995)
587. S. Adachi: J. Appl. Phys. **69**, 7768 (1991)
588. M. Rauh, P. Wißmann, M. Wölfel·
589. G.I. Hornyak, C.J. Patrissi, E.B. Oberhauser, C.R. Martin, J.C. Valmalette, L. Lemaire, J. Dutta, H. Hofmann. Nanostruct. Mat. **9**, 571 (1997)n Solid Films **233**, 289 (1993)
590. J.I. Gittleman, B. Abeles: Phys. Rev. B **15**, 3273 (1977)
591. T. Yamaguchi, M. Sakai, N. Saito: Phys. Rev. B **32**, 2126 (1985)

592. T. Yamaguchi, T. Kitajima, M. Sakai, V.V. Truong: Appl. Surf. Sci. **33/34**, 952 (1988); T. Yamaguchi, R. Chauvaux, V.V. Truong: Physica A **157**, 423 (1989)
593. A.L. Stepanow, D.E. Hole, P.D. Townsend: J. Non.-Crystall. Sol. **244**, 275 (1999)
594. S. Henkel: PhD Thesis, RWTH Aachen (1995)
595. W. Theiss: in *Festkörperprobleme*, Adv. Sol. State Phys. **33** R. Helbig (Ed.), Vieweg Braunschweig (1994) S. 149
596. T. Eickhoff, P. Grosse, S. Henkel, W. Theiss: Z. Phys. B **88**, 17 (1992)
597. M. Evenschor, P. Grosse, W. Theiss: Vibrational Spectrosc. **1**, 173 (1990)
598. J. Sturm, P. Grosse, S. Morley, W. Theiss: Z. Phys. D **26**, S195 (1993)
599. A.S. Barker Jr.: Phys. Rev. B **7**, 2507 (1973)

Printing (Computer to Film): Saladruck Berlin
Binding: Stürtz AG, Würzburg

Springer Series in
MATERIALS SCIENCE

Editors: R. Hull   R. M. Osgood, Jr.   J. Parisi

* The 2nd edition is available as a textbook with the title: *Laser Processing and Chemistry*

Springer Series in
MATERIALS SCIENCE

Editors: R. Hull   R. M. Osgood, Jr.   J. Parisi